Lecture Notes in Statistics

Edited by P. Diggle, S. Fienberg, K. Krickeberg, I. Olkin, N. Wermuth

100

Wlodzimierz Bryc

The Normal Distribution

Characterizations with Applications

Springer-Verlag
New York Berlin Heidelberg London Paris
Tokyo Hong Kong Barcelona Budapest

Wlodzimierz Bryc
Department of Mathematics
University of Cincinnati
PO Box 210025
Cincinnati
Ohio 45221-0025

Library of Congress Cataloging-in-Publication Data Available

© 1995 Springer-Verlag New York, Inc.
All rights reserved. This work may not be translated or copied in whole or in part without the written permission of the publisher (Springer-Verlag New York, Inc., 175 Fifth Avenue, New York, NY 10010, USA), except for brief excerpts in connection with reviews or scholarly analysis. Use in connection with any form of information storage and retrieval, electronic adaptation, computer software, or by similar or dissimilar methodology now known or hereafter developed is forbidden. The use of general descriptive names, trade names, trademarks, etc., in this publication, even if the former are not especially identified, is not to be taken as a sign that such names, as understood by the Trade Marks and Merchandise Marks Act, may accordingly be used freely by anyone.

Camera ready copy provided by the author.

9 8 7 6 5 4 3 2 1

ISBN-13: 978-0-387-97990-8   e-ISBN-13: 978-1-4612-2560-7
DOI: 10.1007/978-1-4612-2560-7

Contents

Preface

This book is a concise presentation of the normal distribution on the real line and its counterparts on more abstract spaces, which we shall call the Gaussian distributions. The material is selected towards presenting *characteristic properties*, or characterizations, of the normal distribution. There are many such properties and there are numerous relevant works in the literature. In this book special attention is given to characterizations generated by the so called Maxwell's Theorem of statistical mechanics, which is stated in the introduction as Theorem 0.0.1. These characterizations are of interest both intrinsically, and as techniques that are worth being aware of. The book may also serve as a good introduction to diverse analytic methods of probability theory. We use characteristic functions, tail estimates, and occasionally dive into complex analysis.

In the book we also show how the characteristic properties can be used to prove important results about the Gaussian processes and the abstract Gaussian vectors. For instance, in Section 5.4 we present Fernique's beautiful proofs of the zero-one law and of the integrability of abstract Gaussian vectors. The central limit theorem is obtained via characterizations in Section 7.3.

The excellent book by Kagan, Linnik & Rao [73] overlaps with ours in the coverage of the classical characterization results. Our presentation of these is sometimes less general, but in return we often give simpler proofs. On the other hand, we are more selective in the choice of characterizations we want to present, and we also point out some applications. Characterization results that are not included in [73] can be found in numerous places of the book, see Section 4.2, Chapter 7 and Chapter 8.

We have tried to make this book accessible to readers with various backgrounds. If possible, we give elementary proofs of important theorems, even if they are special cases of more advanced results. Proofs of several difficult classic results have been simplified. We have managed to avoid functional equations for non-differentiable functions; in many proofs in the literature lack of differentiability is a major technical difficulty.

The book is primarily aimed at graduate students in mathematical statistics and probability theory who would like to expand their bag of tools, to understand the inner workings of the normal distribution, and to explore the connections with other fields. Characterization aspects sometimes show up in unexpected places, cf. Diaconis & Ylvisaker [36]. More generally, when fitting any statistical model to the data, it is inevitable to refer to relevant properties of the population in question; otherwise several different models may fit the same set of empirical data, cf. W. Feller [53]. Monograph [125] by Prakasa Rao is written from such perspective and for a statistician our book may only serve as a complementary source. On the other hand results presented in Sections 7.5 and 8.3 are quite recent and virtually unknown among statisticians. Their modeling aspects remain to be explored, see Section 8.4. We hope that this book will popularize the interesting and difficult area of conditional moment descriptions of random fields. Of course it is possible that such characterizations will finally end up far from real life like many other branches of *applied* mathematics. It is up to the readers of this book to see if the following sentence applies to characterizations as well as to trigonometric series.

> *"Thinking of the extent and refinement reached by the theory of trigonometric series in its long development one sometimes wonders why only relatively few of these advanced achievements find an application."*
> (A. Zygmund, *Trigonometric Series*, Vol. 1, Cambridge Univ. Press, Second Edition, 1959, page xii)

There is more than one way to use this book. Parts of it have been used in a graduate one-quarter course *Topics in statistics*. The reader may also skim through it to find results that he needs; or look up the techniques that might be useful in his own research. The author of this book would be most happy if the reader treats this book as an adventure into the unknown — picks a piece of his liking and follows through and beyond the references. With this is mind, the book has a number of references and digressions. We have tried to point out the historical perspective, but also to get close to current research.

An appropriate background for reading the book is a one year course in real analysis including measure theory and abstract normed spaces, and a one-year course in complex analysis. Familiarity with conditional expectations would also help. Topics from probability theory are reviewed in Chapter 1, frequently with proofs and exercises. Exercise problems are at the end of the chapters; solutions or hints are in Appendix A.

The book benefited from the comments of Chris Burdzy, Abram Kagan, Samuel Kotz, Włodek Smoleński, Paweł Szabłowski, and Jacek Wesołowski. They read portions of the first draft, generously shared their criticism, and pointed out relevant references and errors. My colleagues at the University of Cincinnati also provided comments, criticism and encouragement. The final version of the book was prepared at the Institute for Applied Mathematics of the University of Minnesota in fall quarter of 1993 and at the Center for Stochastic Processes in Chapel Hill in Spring 1994. Support by C. P. Taft Memorial Fund in the summer of 1987 and in the spring of 1994 helped to begin and to conclude this endeavor.

Introduction

The following narrative comes from J. F. W. Herschel [63, page 20].

> *"Suppose a ball dropped from a given height, with the intention that it shall fall on a given mark. Fall as it may, its deviation from the mark is error, and the probability of that error is the unknown function of its square, i. e. of the sum of the squares of its deviations in any two rectangular directions. Now, the probability of any deviation depending solely on its magnitude, and not on its direction, it follows that the probability of each of these rectangular deviations must be the same function of its square. And since the observed oblique deviation is equivalent to the two rectangular ones, supposed concurrent, and which are essentially independent of one another, and is, therefore, a compound event of which they are the simple independent constituents, therefore its probability will be the product of their separate probabilities. Thus the form of our unknown function comes to be determined from this condition..."*

Ten years after Herschel, the reasoning was repeated by J. C. Maxwell [108]. In his theory of gases he assumed that gas consists of small elastic spheres bumping each other; this led to intricate mechanical considerations to analyze the velocities before and after the encounters. However, Maxwell answered the question of his Proposition IV: *What is the distribution of velocities of the gas particles?* without using the details of the interaction between the particles; it lead to the emergence of the trivariate normal distribution. The result that velocities are normally distributed is sometimes called Maxwell's theorem. At the time of discovery, probability theory was in its beginnings and the proof was considered "controversial" by leading mathematicians.

The beauty of the reasoning lies in the fact that the interplay of two very natural assumptions: of independence and of rotation invariance, gives rise to the *normal law of errors* — the most important distribution in statistics. This *interplay of independence and invariance* shows up in many of the theorems presented below.

Here we state the Herschel-Maxwell theorem in modern notation but without proof; for one of the early proofs, see [6]. The reader will see several proofs that use various, usually weaker, assumptions in Theorems 3.1.1, 4.2.1, 5.1.1, 6.3.1, and 6.3.3.

Theorem 0.0.1 *Suppose random variables X, Y have joint probability distribution $\mu(dx, dy)$ such that*

(i) $\mu(\cdot)$ is invariant under the rotations of $\mathbb{R}^2$;

(ii) X, Y are independent.

Then X, Y are normally distributed.

This theorem has generated a vast literature. Here is a quick preview of pertinent results in this book.

Polya's theorem [122] presented in Section 3.1 says that if just two rotations by angles $\pi/2$ and $\pi/4$, preserve the distribution of X, then the distribution is normal. Generalizations to characterizations by the equality of distributions of more general linear forms are given in Chapter 3. One of the most interesting results here is Marcinkiewicz's theorem [106], see Theorem 3.3.3.

An interesting modification of Theorem 0.0.1, discovered by M. Sh. Braverman [14] and presented in Section 4.2 below, considers three i. i. d. random variables X, Y, Z with the *rotation-invariance* assumption (i) replaced by the requirement that only some absolute moments are *rotation invariant*.

Another insight is obtained, if one notices that assumption (i) of Maxwell's theorem implies that rotations preserve the independence of the original random variables X, Y. In this approach we consider a pair X, Y of independent random variables such that the rotation by an angle α produces two independent random variables $X \cos \alpha + Y \sin \alpha$ and $X \sin \alpha - Y \cos \alpha$. Assuming this for all angles α, M. Kac [71] showed that the distribution in question has to be normal. Moreover, careful inspection of Kac's proof reveals that the only essential property he had used was that X, Y are independent and that just one $\pi/4$-rotation: $(X+Y)/\sqrt{2}, (X-Y)/\sqrt{2}$ produces the independent pair. The result explicitly assuming the latter was found independently by Bernstein [8]. Bernstein's theorem and its extensions are considered in Chapter 5; Bernstein's theorem also motivates the assumptions in Chapter 7.

The following is a more technical description the contents of the book. Chapter 1 collects probabilistic prerequisites. The emphasis is on analytic aspects; in particular elementary but useful tail estimates collected in Section 1.3. In Chapter 2 we approach multivariate normal distributions through characteristic functions. This is a less intuitive but powerful method. It leads rapidly to several fundamental facts, and to associated Reproducing Kernel Hilbert Spaces (RKHS). As an illustration, we prove the large deviation estimates on $\mathbb{R}^d$ which use the conjugate RKHS norm. In Chapter 3 the reader is introduced to stability and equidistribution of linear forms in independent random variables. Stability is directly related to the CLT. We show that in the abstract setup stability is also responsible for the zero-one law. Chapter 4 presents the analysis of rotation invariant distributions on $\mathbb{R}^d$ and on $\mathbb{R}^\infty$. We study when a rotation invariant distribution has to be normal. In the process we analyze structural properties of rotation invariant laws and introduce the relevant techniques. In this chapter we also present surprising results on rotation invariance of the absolute moments. We conclude with a short proof of de Finetti's theorem and point out its implications for infinite spherically symmetric sequences. Chapter 5 parallels Chapter 3 in analyzing the role of independence of linear forms. We show that independence of certain linear forms, a characteristic property of the normal distribution, leads to the zero-one law, and it is also responsible for exponential moments. Chapter 6 is a short introduction to measures of dependence and stability issues. Theorem 6.2.2 establishes integrability under conditions of interest, eg. in *polynomial biorthogonality* as studied by Lancaster [94]. In Chapter 7 we extend results in Chapter 5 to conditional moments. Three interesting aspects emerge here. First, normality can frequently be recognized from the conditional moments of linear combinations of independent random variables; we illustrate this by a simple proof of the well known fact that the independence of the sample mean and the sample variance characterizes normal populations, and by the proof of the central limit theorem. Secondly, we show that for infinite sequences, conditional moments determine normality without any reference to independence. This part has its natural continuation in Chapter 8. Thirdly, in the exercises we point out the versatility of conditional moments in handling other infinitely divisible distributions. Chapter 8 is a short introduction to continuous parameter random fields, analyzed through their conditional moments. We also present a self-contained analytic construction of the Wiener process.

Chapter 1

Probability tools

Most of the contents of this section is fairly standard probability theory. The reader shouldn't be under the impression that this chapter is a substitute for a systematic course in probability theory. We will skip important topics such as limit theorems. The emphasis here is on analytic methods; in particular characteristic functions will be extensively used throughout.

Let $(\Omega, \mathcal{M}, P)$ be the probability space, ie. Ω is a set, $\mathcal{M}$ is a σ-field of its subsets and P is the probability measure on $(\Omega, \mathcal{M})$. We follow the usual conventions: X, Y, Z stand for real random variables; boldface $\mathbf{X}, \mathbf{Y}, \mathbf{Z}$ denote vector-valued random variables. Throughout the book $EX = \int_{\Omega} X(\omega)\, dP$ (Lebesgue integral) denotes the expected value of a random variable X. We write $X \cong Y$ to denote the equality of distributions, ie. $P(X \in A) = P(Y \in A)$ for all measurable sets A. Equalities and inequalities between random variables are to be interpreted almost surely (a. s.). For instance $X \leq Y + 1$ means $P(X \leq Y + 1) = 1$; the latter is a shortcut that we use for the expression $P(\{\omega \in \Omega : X(\omega) \leq Y(\omega) + 1\}) = 1$.

Boldface $\mathbf{A}, \mathbf{B}, \mathbf{C}$ will denote matrices. For a complex $z = x + iy \in \mathbf{C}$ by $x = \Re z$ and $y = \Im z$ we denote the real and the imaginary part of z. Unless otherwise stated, $\log a = \log_e a$ denotes the natural logarithm of number a.

1.1 Moments

Given a real number $r \geq 0$, the absolute moment of order r is defined by $E|X|^r$; the ordinary moment of order $r = 0, 1, \ldots$ is defined as EX^r. Clearly, not every sequence of numbers is the sequence of moments of a random variable X; it may also happen that two random variables with different distributions have the same moments. However, in Corollary 2.3.3 below we will show that the latter cannot happen for normal distributions.

The following inequality is known as Chebyshev's inequality. Despite its simplicity it has numerous non-trivial applications, see eg. Theorem 6.2.2 or [29].

Proposition 1.1.1 *If $f : \mathbb{R}_+ \to \mathbb{R}_+$ is a non-decreasing function and $Ef(|X|) = C < \infty$, then for all $t > 0$ such that $f(t) \neq 0$ we have*

$$P(|X| > t) \leq C/f(t). \tag{1.1}$$

Indeed, $Ef(|X|) = \int_\Omega f(|X|)\,dP \geq \int_{|X|\geq t} f(|X|)\,dP \geq \int_{|X|\geq t} f(t)\,dP = f(t)P(|X| > t)$.

It follows immediately from Chebyshev's inequality that if $E|X|^p = C < \infty$, then $P(|X| > t) \leq C/t^p, t > 0$. An implication in converse direction is also well known: if $P(|X| > t) \leq C/t^{p+\epsilon}$ for some $\epsilon > 0$ and for all $t > 0$, then $E|X|^p < \infty$, see (1.4) below.

The following formula will often be useful[1].

Proposition 1.1.2 *If* $f : \mathbb{R}_+ \rightarrow \mathbb{R}$ *is a function such that* $f(x) = f(0) + \int_0^x g(t)\,dt$, $E\{|f(X)|\} < \infty$ *and* $X \geq 0$, *then*

$$Ef(X) = f(0) + \int_0^\infty g(t)P(X \geq t)\,dt. \tag{1.2}$$

Moreover, if $g \geq 0$ *and if the right hand side of (1.2) is finite, then* $Ef(X) < \infty$.

Proof. The formula follows from Fubini's theorem[2], since for $X \geq 0$

$$\int_\Omega f(X)\,dP = \int_\Omega \left(f(0) + \int_0^\infty 1_{t\leq X} g(t)\,dt \right) dP$$

$$= f(0) + \int_0^\infty g(t)\left(\int_\Omega 1_{t\leq X}\,dP\right)dt = f(0) + \int_0^\infty g(t)P(X \geq t)\,dt.$$

$\square$

Corollary 1.1.3 *If* $E|X|^r < \infty$ *for an integer* $r > 0$, *then*

$$EX^r = r\int_0^\infty t^{r-1}P(X \geq t)\,dt - r\int_0^\infty t^{r-1}P(-X \geq t)\,dt. \tag{1.3}$$

If $E|X|^r < \infty$ *for real* $r > 0$ *then*

$$E|X|^r = r\int_0^\infty t^{r-1}P(|X| \geq t)\,dt. \tag{1.4}$$

Moreover, the left hand side of (1.4) is finite if and only if the right hand side is finite.

Proof. Formula (1.4) follows directly from Proposition 1.1.2 (with $f(x) = x^r$ and $g(t) = \frac{d}{dt}f(t) = rt^{r-1}$).

Since $EX = EX^+ - EX^-$, where $X^+ = \max\{X, 0\}$ and $X^- = \min\{X, 0\}$, therefore applying Proposition 1.1.2 separately to each of this expectations we get (1.3). $\square$

---

[1] The typical application deals with $Ef(X)$ when $f(.)$ has continuous derivative, or when $f(.)$ is convex. Then the integral representation from the assumption holds true.

[2] See, eg. [9, Section 18] .

1.2 L_p-spaces

By $L_p(\Omega, \mathcal{M}, P)$, or L_p if no misunderstanding may result, we denote the Banach space of a. s. classes of equivalence of p-integrable $\mathcal{M}$-measurable random variables X with the norm

$$\|X\|_p = \begin{cases} \sqrt[p]{E|X|^p} & \text{if } p \geq 1; \\ \text{ess sup}|X| & \text{if } p = \infty. \end{cases}$$

If $X \in L_p$, we shall say that X is p-integrable; in particular, X is square integrable if $EX^2 < \infty$. We say that X_n converges to X in L_p, if $\|X_n - X\|_p \to 0$ as $n \to \infty$. If X_n converges to X in L_2, we shall also use the phrase *sequence X_n converges to X in mean-square*.

Several useful inequalities are collected in the following.

Theorem 1.2.1   *(i) for $1 \leq p \leq q \leq \infty$ we have Minkowski's inequality*

$$\|X\|_p \leq \|X\|_q. \tag{1.5}$$

(ii) for $1/p + 1/q = 1$, $p \geq 1$ we have Hölder's inequality

$$EXY \leq \|X\|_p \|Y\|_q. \tag{1.6}$$

(iii) for $1 \leq p \leq \infty$ we have triangle inequality

$$\|X + Y\|_p \leq \|X\|_p + \|Y\|_p. \tag{1.7}$$

Special case $p = q = 2$ of Hölder's inequality (1.6) reads $EXY \leq \sqrt{EX^2 EY^2}$. It is frequently used and is known as the *Cauchy-Schwarz inequality*.

For $1 \leq p < \infty$ the conjugate space to L_p (ie. the space of all bounded linear functionals on L_p) is usually identified with L_q, where $1/p + 1/q = 1$. The identification is by the duality $\langle f, g \rangle = \int f(\omega) g(\omega) \, dP$.

For the proof of Theorem 1.2.1 we need the following elementary inequality.

Lemma 1.2.2 *For $a, b > 0, 1 < p < \infty$ and $1/p + 1/q = 1$ we have*

$$ab \leq a^p/p + b^q/q. \tag{1.8}$$

Proof. Function $t \mapsto t^p/p + t^{-q}/q$ has the derivative $t^{p-1} - t^{-q-1}$. The derivative is positive for $t > 1$ and negative for $0 < t < 1$. Hence the maximum value of the function for $t > 0$ is attained at $t = 1$, giving

$$t^p/p + t^{-q}/q \geq 1.$$

Substituting $t = a^{1/q} b^{-1/p}$ we get (1.8). $\square$

Proof of Theorem 1.2.1 (ii). If either $\|X\|_p = 0$ or $\|Y\|_q = 0$, then $XY = 0$ a. s. Therefore we consider only the case $\|X\|_p \|Y\|_q > 0$ and after rescaling we assume $\|X\|_p =$

$\|Y\|_q = 1$. Furthermore, the case $p = 1, q = \infty$ is trivial as $|XY| \leq |X|\|Y\|_\infty$. For $1 < p < \infty$ by (1.8) we have

$$|XY| \leq |X|^p/p + |Y|^q/q.$$

Integrating this inequality we get $|EXY| \leq E|XY| \leq 1 = \|X\|_p\|Y\|_q$. $\square$

Proof of Theorem 1.2.1 (i). For $p = 1$ this is just Jensen's inequality; for a more general version see Theorem 1.4.1. For $1 < p < \infty$ by Hölder's inequality applied to the product of 1 and $|X|^p$ we have

$$\|X\|_p^p = E\{|X|^p \cdot 1\} \leq (E|X|^q)^{p/q}(E1^r)^{1/r} = \|X\|_q^p,$$

where r is computed from the equation $1/r + p/q = 1$. (This proof works also for $p = 1$ with obvious changes in the write-up.) $\square$

Proof of Theorem 1.2.1 (iii). The inequality is trivial if $p = 1$ or if $\|X + Y\|_p = 0$. In the remaining cases

$$\|X + Y\|_p^p \leq E\{(|X| + |Y|)|X + Y|^{p-1}\} = E\{|X||X + Y|^{p-1}\} + E\{|Y||X + Y|^{p-1}\}.$$

By Hölder's inequality

$$\|X + Y\|_p^p \leq \|X\|_p\|X + Y\|_p^{p/q} + \|Y\|_p\|X + Y\|_p^{p/q}.$$

Since $p/q = p - 1$, dividing both sides by $\|X + Y\|_p^{p/q}$ we get the conclusion. $\square$

By $Var(X)$ we shall denote the variance of a square integrable r. v. X

$$Var(X) = EX^2 - (EX)^2 = E(X - EX)^2.$$

The correlation coefficient $corr(X, Y)$ is defined for square-integrable non-degenerate r. v. X, Y by the formula

$$corr(X, Y) = \frac{EXY - EXEY}{\|X - EX\|_2\|Y - EY\|_2}.$$

The Cauchy-Schwarz inequality implies that $-1 \leq corr(X, Y) \leq 1$.

1.3 Tail estimates

The function $N(x) = P(|X| \geq x)$ describes *tail behavior* of r. v. a X. Inequalities involving $N(\cdot)$ similar to Problems 1.2 and 1.3 are sometimes easy to prove. Integrability that follows is of considerable interest. Below we give two rather technical tail estimates and we state several corollaries for future reference. The proofs use only the fact that $N : [0, \infty) \to [0, 1]$ is a non-increasing function such that $\lim_{x \to \infty} N(x) = 0$.

Theorem 1.3.1 *If there are $C > 1, 0 < q < 1, x_0 \geq 0$ such that for all $x > x_0$*

$$N(Cx) \leq qN(x - x_0), \tag{1.9}$$

then there is $M < \infty$ such that $N(x) \leq \frac{M}{x^\beta}$, where $\beta = -\log_C q$.

Proof. Let a_n be such that when $a_n = x_n - x_0$ then $a_{n+1} = Cx_n$. Solving the resulting recurrence we get $a_n = C^n - b$, where $b = Cx_0(C - 1)^{-1}$. Equation (1.9) says $N(a_{n+1}) \leq CN(a_n)$. Therefore

$$N(a_n) \leq N(a_0)q^n.$$

This implies the tail estimate for arbitrary $x > 0$. Namely, given $x > 0$ choose n such that $a_n \leq x < a_{n+1}$. Then

$$N(x) \leq N(a_n) \leq Kq^n = \frac{K}{q}q^{\log_C(a_{n+1}+b)} = M(x + b)^{-\beta}.$$

$\square$

The next results follow from Theorem 1.3.1 and (1.4) and are stated for future reference.

Corollary 1.3.2 *If there is $0 < q < 1$ and $x_0 \geq 0$ such that $N(2x) \leq qN(x - x_0)$ for all $x > x_0$, then $E|X|^\beta < \infty$ for all $\beta < \log_2 1/q$.*

Corollary 1.3.3 *Suppose there is $C > 1$ such that for every $0 < q < 1$ one can find $x_0 \geq 0$ such that*

$$N(Cx) \leq qN(x) \tag{1.10}$$

for all $x > x_0$. Then $E|X|^p < \infty$ for all p.

As a special case of Corollary 1.3.3 we have the following.

Corollary 1.3.4 *Suppose there are $C > 1, K < \infty$ such that*

$$N(Cx) \leq K\frac{N(x)}{x^2} \tag{1.11}$$

for all x large enough. Then $E|X|^p < \infty$ for all p.

The next result deals with exponentially small tails.

Theorem 1.3.5 *If there are $C > 1, 1 < K < \infty, x_0 \geq 0$ such that*

$$N(Cx) \leq KN^2(x - x_0) \tag{1.12}$$

for all $x > x_0$, then there are $M < \infty, \beta > 0$ such that

$$N(x) \leq M\exp(-\beta x^\alpha),$$

where $\alpha = \log_C 2$.

Proof. As in the proof of Theorem 1.3.1, let $a_n = C^n - b, b = Cx_0/(C-1)$. Put $q_n = \log_K N(a_n)$. Then (1.12) gives

$$N(a_{n+1}) \leq K N^2(a_n),$$

which implies

$$q_{n+1} \leq 2q_n + 1. \tag{1.13}$$

Therefore by induction we get

$$q_{m+n} \leq 2^n(1 + q_m) - 1. \tag{1.14}$$

Indeed, (1.14) becomes equality for $n = 0$. If it holds for $n = k$, then $q_{m+k+1} \leq 2q_{m+k}+1 \leq 2(2^k(1 + q_m) - 1) + 1 = 2^{k+1}(1 + q_m) - 1$. This proves (1.14) by induction.

Since $a_n \to \infty$, we have $N(a_n) \to 0$ and $q_n \to -\infty$. Choose m large enough to have $1 + q_m < 0$. Then (1.14) implies

$$N(a_{n+m}) \leq K^{2^n(1+q_m)} = \exp{-\beta 2^n}.$$

The proof is now concluded by the standard argument. Selecting large enough M we have $N(x) \leq 1 \leq M \exp{-\beta x^\alpha}$ for all $x \leq a_m$. Given $x > a_m$ choose $n \geq 0$ such that $a_{n+m} \leq x < a_{n+m+1}$. Then

$$N(x) \leq N(a_{n+m}) \leq \exp{-\beta 2^n} \leq M \exp(-\beta 2^{\log_C a_{n+m+1}}) \leq M \exp{-\beta x^\alpha}.$$

$\square$

Corollary 1.3.6 *If there are $C < \infty, x_0 \geq 0$ such that*

$$N(\sqrt{2}x) \leq C N^2(x - x_0),$$

then there is $\beta > 0$ such that $E \exp(\beta|X|^2) < \infty$.

Corollary 1.3.7 *If there are $C < \infty, x_0 \geq 0$ such that*

$$N(2x) \leq C N^2(x - x_0),$$

then there is $\beta > 0$ such that $E \exp(\beta|X|) < \infty$.

1.4   Conditional expectations

Below we recall the definition of the conditional expectation of a r. v. with respect to a σ-field and we state several results that we need for future reference. The definition is as old as axiomatic probability theory itself, see [82, Chapter V page 53 formula (2)]. The reader not familiar with conditional expectations should consult textbooks, eg. Billingsley [9, Section 34], Durrett [42, Chapter 4], or Neveu [117].

Definition 1.4.1 *Let $(\Omega, \mathcal{M}, P)$ be a probability space. If $\mathcal{F} \subset \mathcal{M}$ is a σ-field and X is an integrable random variable, then the conditional expectation of X given $\mathcal{F}$ is an integrable $\mathcal{F}$-measurable random variable Z such that $\int_A X\, dP = \int_A Z\, dP$ for all $A \in \mathcal{F}$.*

Conditional expectation of an integrable random variable X with respect to a σ-field $\mathcal{F} \subset \mathcal{M}$ will be denoted interchangeably by $E\{X|\mathcal{F}\}$ and $E^{\mathcal{F}}X$. We shall also write $E\{X|Y\}$ or $E^Y X$ for the conditional expectation $E\{X|\mathcal{F}\}$ when $\mathcal{F} = \sigma(Y)$ is the σ-field generated by a random variable Y.

Existence and almost sure uniqueness of the conditional expectation $E\{X|\mathcal{F}\}$ follows from the Radon-Nikodym theorem, applied to the finite signed measures $\mu(A) = \int_A X\,dP$ and $P_{|\mathcal{F}}$, both defined on the measurable space $(\Omega, \mathcal{F})$. In some simple situations more explicit expressions can also be found.

Example. *Suppose $\mathcal{F}$ is a σ-field generated by the events $A_1, A_2, \ldots, A_n$ which form a non-degenerate disjoint partition of the probability space Ω. Then it is easy to check that*

$$E\{X|\mathcal{F}\}(\omega) = \sum_{k=1}^n m_k I_{A_k}(\omega),$$

where $m_k = \int_{A_k} X\,dP/P(A_k)$. In other words, on A_k we have $E\{X|\mathcal{F}\} = \int_{A_k} X\,dP/P(A_k)$. In particular, if X is discrete and $X = \sum x_j I_{B_j}$, then we get intuitive expression

$$E\{X|\mathcal{F}\} = \sum x_j P(B_j|A_k) \text{ for } \omega \in A_k.$$

Example. *Suppose that $f(x,y)$ is the joint density with respect to the Lebesgue measure on $\mathbb{R}^2$ of the bivariate random variable (X,Y) and let $f_Y(y) \neq 0$ be the (marginal) density of Y. Put $f(x|y) = f(x,y)/f_Y(y)$. Then $E\{X|Y\} = h(Y)$, where $h(y) = \int_{-\infty}^{\infty} x f(x|y)\,dx$.*

The next theorem lists properties of conditional expectations that will be used without further mention.

Theorem 1.4.1   *(i) If Y is $\mathcal{F}$-measurable random variable such that X and XY are integrable, then $E\{XY|\mathcal{F}\} = YE\{X|\mathcal{F}\}$;*

(ii) If $\mathcal{G} \subset \mathcal{F}$, then $E^{\mathcal{G}}E^{\mathcal{F}} = E^{\mathcal{G}}$;

(iii) If $\sigma(X, \mathcal{F})$ and $\mathcal{N}$ are independent σ-fields, then $E\{X|\mathcal{N}\vee\mathcal{F}\} = E\{X|\mathcal{F}\}$; here $\mathcal{N}\vee\mathcal{F}$ denotes the σ-field generated by the union $\mathcal{N}\cup\mathcal{F}$;

(iv) If $g(x)$ is a convex function and $E|g(X)| < \infty$, then $g(E\{X|\mathcal{F}\}) \leq E\{g(X)|\mathcal{F}\}$;

(v) If $\mathcal{F}$ is the trivial σ-field consisting of the events of probability 0 or 1 only, then $E\{X|\mathcal{F}\} = EX$;

(vi) If X, Y are integrable and $a, b \in \mathbb{R}$ then $E\{aX + bY|\mathcal{F}\} = aE\{X|\mathcal{F}\} + bE\{Y|\mathcal{F}\}$;

(vii) If X and $\mathcal{F}$ are independent, then $E\{X|\mathcal{F}\} = EX$.

Remark: Inequality (iv) is known as Jensen's inequality and this is how we shall refer to it.

The proof uses the following.

Lemma 1.4.2 *If Y_1 and Y_2 are $\mathcal{F}$-measurable and $\int_A Y_1\,dP \leq \int_A Y_2\,dP$ for all $A \in \mathcal{F}$, then $Y_1 \leq Y_2$ almost surely. If $\int_A Y_1\,dP = \int_A Y_2\,dP$ for all $A \in \mathcal{F}$, then $Y_1 = Y_2$.*

Proof. Let $A_\epsilon = \{Y_1 > Y_2 + \epsilon\} \in \mathcal{F}$. Since $\int_{A_\epsilon} Y_1\, dP \geq \int_{A_\epsilon} Y_2\, dP + \epsilon P(A_\epsilon)$, thus $P(A_\epsilon) > 0$ is impossible. Event $\{Y_1 > Y_2\}$ is the countable union of the events A_ϵ (with ϵ rational); thus it has probability 0 and $Y_1 \leq Y_2$ with probability one.

The second part follows from the first by symmetry. $\square$

Proof of Theorem 1.4.1.

(i) This is verified first for $Y = I_B$ (the indicator function of an event $B \in \mathcal{F}$). Let $Y_1 = E\{XY|\mathcal{F}\}, Y_2 = Y E\{X|\mathcal{F}\}$. From the definition one can easily see that both $\int_A Y_1\, dP$ and $\int_A Y_2\, dP$ are equal to $\int_{A \cap B} X\, dP$. Therefore $Y_1 = Y_2$ by the Lemma 1.4.2.

For the general case, approximate Y by simple random variables and use (vi).

(ii) This follows from Lemma 1.4.2: random variables $Y_1 = E\{X|\mathcal{F}\}, Y_2 = E\{X|\mathcal{G}\}$ are $\mathcal{G}$-measurable and for A in $\mathcal{G}$ both $\int_A Y_1\, dP$ and $\int_A Y_2\, dP$ are equal to $\int_A X\, dP$.

(iii) Let $Y_1 = E\{X|\mathcal{N} \vee \mathcal{F}\}, Y_2 = E\{X|\mathcal{F}\}$. We check first that

$$\int_A Y_1\, dP = \int_A Y_2\, dP$$

for all $A = B \cap C$, where $B \in \mathcal{N}$ and $C \in \mathcal{F}$. This holds true, as both sides of the equation are equal to $P(B) \int_C X\, dP$. Once equality $\int_A Y_1\, dP = \int_A Y_2\, dP$ is established for the generators of the σ-field, it holds true for the whole σ-field $\mathcal{N} \vee \mathcal{F}$; this is standard measure theory, see $\pi - \lambda$ Theorem [9, Theorem 3.3].

(iv) Here we need the first part of Lemma 1.4.2. We also need to know that each convex function $g(x)$ can be written as the supremum of a family of affine functions $f_{a,b}(x) = ax + b$. Let $Y_1 = E\{g(X)|\mathcal{F}\}, Y_2 = f_{a,b}(E\{X|\mathcal{F}\}), A \in \mathcal{F}$. By (vi) we have

$$\int_A Y_1\, dP = \int_A g(X)\, dP \geq f_{a,b}\!\left(\int_A X\right) dP = f_{a,b}\!\left(\int_A E\{X|\mathcal{F}\}\right) dP = \int_A Y_2\, dP.$$

Hence $f_{a,b}(E\{X|\mathcal{F}\}) \leq E\{g(X)|\mathcal{F}\}$; taking the supremum (over suitable a, b) ends the proof.

(v), (vi), (vii) These proofs are left as exercises. $\square$

Theorem 1.4.1 gives geometric interpretation of the conditional expectation $E\{\cdot|\mathcal{F}\}$ as the projection of the Banach space $L_p(\Omega, \mathcal{M}, P)$ onto its closed subspace $L_p(\Omega, \mathcal{F}, P)$, consisting of all p-integrable $\mathcal{F}$-measurable random variables, $p \geq 1$. This projection is "self adjoint" in the sense that the adjoint operator is given by the same "conditional expectation" formula, although the adjoint operator acts on L_q rather than on L_p; for square integrable functions $E\{.|\mathcal{F}\}$ is just the orthogonal projection onto $L_2(\Omega, \mathcal{F}, P)$. Monograph [117] considers conditional expectation from this angle.

We will use the following (weak) version of the martingale[3] convergence theorem.

Theorem 1.4.3 *Suppose $\mathcal{F}_n$ is a decreasing family of σ-fields, ie. $\mathcal{F}_{n+1} \subset \mathcal{F}_n$ for all $n \geq 1$. If X is integrable, then $E\{X|\mathcal{F}_n\} \to E\{X|\mathcal{F}\}$ in L_1-norm, where $\mathcal{F}$ is the intersection of all $\mathcal{F}_n$.*

---

[3]A martingale with respect to a family of increasing σ-fields $\mathcal{F}_n$ is and integrable sequence X_n such that $E(X_{n+1}|\mathcal{F}_n) = X_n$. The sequence $X_n = E(X|\mathcal{F}_n)$ is a martingale. The sequence in the theorem is of the same form, except that the σ-fields are decreasing rather than increasing.

Proof. Suppose first that X is square integrable. Subtracting $m = EX$ if necessary, we can reduce the convergence question to the centered case $EX = 0$. Denote $X_n = E\{X|\mathcal{F}_n\}$. Since $\mathcal{F}_{n+1} \subset \mathcal{F}_n$, by Jensen's inequality $EX_n^2 \geq 0$ is a decreasing non-negative sequence. In particular, EX_n^2 converges.

Let $m < n$ be fixed. Then $E(X_n - X_m)^2 = EX_n^2 + EX_m^2 - 2EX_nX_m$. Since $\mathcal{F}_n \subset \mathcal{F}_m$, by Theorem 1.4.1 we have

$$EX_nX_m = EE\{X_nX_m|\mathcal{F}_n\} = EX_nE\{X_m|\mathcal{F}_n\}$$

$$= EX_nE\{E\{X|\mathcal{F}_m\}|\mathcal{F}_n\} = EX_nE\{X|\mathcal{F}_n\} = EX_n^2.$$

Therefore $E(X_n - X_m)^2 = EX_m^2 - EX_n^2$. Since EX_n^2 converges, X_n satisfies the Cauchy condition for convergence in L_2 norm. This shows that for square integrable X, sequence $\{X_n\}$ converges in L_2.

If X is not square integrable, then for every $\epsilon > 0$ there is a square integrable Y such that $E|X - Y| < \epsilon$. By Jensen's inequality $E\{X|\mathcal{F}_n\}$ and $E\{Y|\mathcal{F}_n\}$ differ by at most ϵ in L_1-norm; this holds uniformly in n. Since by the first part of the proof $E\{Y|\mathcal{F}_n\}$ is convergent, it satisfies the Cauchy condition in L_2 and hence in L_1. Therefore for each $\epsilon > 0$ we can find N such that for all $n, m > N$ we have $E\{|E\{X|\mathcal{F}_n\} - E\{X|\mathcal{F}_m\}|\} < 3\epsilon$. This shows that $E\{X|\mathcal{F}_n\}$ satisfies the Cauchy condition and hence converges in L_1.

The fact that the limit is $X_\infty = E\{X|\mathcal{F}\}$ can be seen as follows. Clearly X_∞ is $\mathcal{F}_n$-measurable for all n, ie. it is $\mathcal{F}$-measurable. For $A \in \mathcal{F}$ (hence also in $\mathcal{F}_n$), we have $EXI_A = EX_nI_A$. Since $|EX_nI_A - EX_\infty I_A| \leq E|X_n - X_\infty|I_A \leq E|X_n - X_\infty| \to 0$, therefore $EX_nI_A \to EX_\infty I_A$. This shows that $EXI_A = EX_\infty I_A$ and by definition, $X_\infty = E\{X|\mathcal{F}\}$. $\square$

1.5 Characteristic functions

The characteristic function of a real-valued random variable X is defined by $\phi_X(t) = E\exp(itX)$, where i is the imaginary unit ($i^2 = -1$). It is easily seen that

$$\phi_{aX+b}(t) = e^{itb}\phi_X(at). \tag{1.15}$$

If X has the density $f(x)$, the characteristic function is just its Fourier transform: $\phi(t) = \int_{-\infty}^{\infty} e^{itx}f(x)\,dx$. If $\phi(t)$ is integrable, then the inverse Fourier transform gives

$$f(x) = \frac{1}{2\pi} \int_{-\infty}^{\infty} e^{-itx}\phi(t)\,dt.$$

This is occasionally useful in verifying whether the specific $\phi(t)$ is a characteristic function as in the following example.

Example 1.5.1 *The following gives an example of characteristic function that has finite support. Let $\phi(t) = 1 - |t|$ for $|t| < |\ < 1$ and 0 otherwise. Then*

$$f(x) = \frac{1}{2\pi} \int_{-1}^{1} e^{-itx}(1 - |t|)\,dt = -\frac{1}{\pi} \int_{0}^{1}(1 - t)\cos tx\,dt = \frac{1}{\pi}\frac{1 - \cos x}{x^2}.$$

Since $f(x) = \frac{1}{\pi}\frac{1-\cos x}{x^2}$ is non-negative and integrable, $\phi(t)$ is indeed a characteristic function.

The following properties of characteristic functions are proved in any standard probability course, see eg. [9, 54].

Theorem 1.5.1 *(i) The distribution of X is determined uniquely by its characteristic function $\phi(t)$.*

(ii) If $E|X|^r < \infty$ for some $r = 0, 1, \ldots$, then $\phi(t)$ is r-times differentiable, the derivative is uniformly continuous and

$$EX^k = (-i)^k \frac{d^k}{dt^k}\phi(t)\bigg|_{t=0}$$

for all $0 \le k \le r$.

(iii) If $\phi(t)$ is 2r-times differentiable for some natural r, then $EX^{2r} < \infty$.

(iv) If X, Y are independent random variables, then $\phi_{X+Y}(t) = \phi_X(t)\phi_Y(t)$ for all $t \in \mathbb{R}$.

For a d-dimensional random variable $\mathbf{X} = (X_1, \ldots, X_d)$ the characteristic function $\phi_{\mathbf{X}} : \mathbb{R}^d \to \mathbb{C}$ is defined by $\phi_{\mathbf{X}}(\mathbf{t}) = E\exp(i t \cdot \mathbf{X})$, where the dot denotes the dot (scalar) product, ie. $\mathbf{x} \cdot \mathbf{y} = \sum x_k y_k$. For a pair of real valued random variables X, Y, we also write $\phi(t, s) = \phi_{(X,Y)}((t, s))$ and we call $\phi(t, s)$ the joint characteristic function of X and Y.

The following is the multi-dimensional version of Theorem 1.5.1.

Theorem 1.5.2 *(i) The distribution of $\mathbf{X}$ is determined uniquely by its characteristic function $\phi(\mathbf{t})$.*

(ii) If $E\|\mathbf{X}\|^r < \infty$, then $\phi(\mathbf{t})$ is r-times differentiable and

$$EX_{j_1} \ldots X_{j_k} = (-i)^k \frac{\partial^k}{\partial t_{j_1} \ldots \partial t_{j_k}}\phi(\mathbf{t})\bigg|_{\mathbf{t}=0}$$

for all $0 \le k \le r$.

(iii) If $\mathbf{X}, \mathbf{Y}$ are independent $\mathbb{R}^d$-valued random variables, then

$$\phi_{\mathbf{X}+\mathbf{Y}}(\mathbf{t}) = \phi_{\mathbf{X}}(\mathbf{t})\phi_{\mathbf{Y}}(\mathbf{t})$$

for all $\mathbf{t}$ in $\mathbb{R}^d$.

The next result seems to be less known although it is both easy to prove and to apply. We shall use it on several occasions in Chapter 7. The converse is also true if we assume that the integer parameter r in the proof below is even or that joint characteristic function $\phi(t, s)$ is differentiable; to prove the converse, one can follow the usual proof of the inversion formula for characteristic functions, see, eg. [9, Section 26]. Kagan, Linnik & Rao [73, Section 1.1.5] state explicitly several most frequently used variants of (1.17).

Theorem 1.5.3 *Suppose real valued random variables X, Y have the joint characteristic function $\phi(t, s)$. Assume that $E|X|^m < \infty$ for some $m \in \mathbb{N}$. Let $g(y)$ be such that*

$$E\{X^m|Y\} = g(Y).$$

Then for all real s

$$(-i)^m \frac{\partial^m}{\partial t^m} \phi(t,s)\bigg|_{t=0} = E g(Y) \exp(isY). \qquad (1.16)$$

In particular, if $g(y) = \sum c_k y^k$ is a polynomial, then

$$(-i)^m \frac{\partial^m}{\partial t^m} \phi(t,s)\bigg|_{t=0} = \sum_k (-i)^k c_k \frac{d^k}{ds^k} \phi(0,s). \qquad (1.17)$$

Proof. Since by assumption $E|X|^m < \infty$, the joint characteristic function $\phi(t,s) = E\exp(itX + isY)$ can be differentiated m times with respect to t and

$$\frac{\partial^m}{\partial t^m} \phi(t,s) = i^m E X^m \exp(itX + isY).$$

Putting t=0 establishes (1.16), see Theorem 1.4.1(i).

In order to prove (1.17), we need to show first that $E|Y|^r < \infty$, where r is the degree of the polynomial $g(y)$. By Jensen's inequality $E|g(Y)| \le E|X|^m < \infty$, and since $|g(y)/y^r| \to const \ne 0$ as $|y| \to \infty$, therefore there is $C > 0$ such that $|y|^r \le C|g(y)|$ for all y. Hence $E|Y|^r < \infty$ follows.

Formula (1.17) is now a simple consequence of (1.16); indeed, for $0 \le k \le r$ we have $EY^k \exp(isY) = (-i)^k k \phi(0,s)$; this formula is obtained by differentiating k-times $E\exp(isY)$ under the integral sign. $\square$

1.6 Symmetrization

Definition 1.6.1 *A random variable X (also: a vector valued random variable $\mathbf{X}$) is symmetric if X and $-X$ have the same distribution.*

Symmetrization techniques deal with comparison of properties of an arbitrary variable X with some symmetric variable X_{sym}. Symmetric variables are usually easier to deal with, and proofs of many theorems (not only characterization theorems, see eg. [76]) become simpler when reduced to the symmetric case.

There are two natural ways to obtain a symmetric random variable X_{sym} from an arbitrary random variable X. The first one is to multiply X by an independent random sign ± 1; in terms of the characteristic functions this amounts to replacing the characteristic function ϕ of X by its symmetrization $\frac{1}{2}(\phi(t) + \phi(-t))$. This approach has the advantage that if X is symmetric, then its symmetrization X_{sym} has the same distribution as X. Integrability properties are also easy to compare, because $|X| = |X_{sym}|$.

The other symmetrization, which has perhaps less obvious properties but is frequently found more useful, is defined as follows. Let X' be an independent copy of X. The symmetrization $\widetilde{X}$ of X is defined by $\widetilde{X} = X - X'$. In terms of the characteristic functions this corresponds to replacing the characteristic function $\phi(t)$ of X by the characteristic function $|\phi(t)|^2$. This procedure is easily seen to change the distribution of X, except when $X = 0$.

Theorem 1.6.1 *(i) If the symmetrization $\widetilde{X}$ of a random variable X has a finite moment of order $p \geq 1$, then $E|X|^p < \infty$.*

(ii) If the symmetrization $\widetilde{X}$ of a random variable X has finite exponential moment $E\exp(\lambda|\widetilde{X}|)$, then $E\exp\lambda|X| < \infty, \lambda > 0$.

(iii) If the symmetrization $\widetilde{X}$ of a random variable X satisfies $E\exp\lambda|\widetilde{X}|^2 < \infty$, then $E\exp\lambda|X|^2 < \infty, \lambda > 0$.

The usual approach to Theorem 1.6.1 uses the symmetrization inequality, which is of independent interest (see Problem 1.20) and formula (1.2). Our proof requires extra assumptions, but instead is short, does not require X and X' to have the same distribution, and it also gives a more accurate bound (within its domain of applicability).

Proof in the case, when $E|X| < \infty$ and $EX = 0$: Let $g(x) \geq 0$ be a convex function, such that $Eg(\widetilde{X}) < \infty$ and let X, X' be the independent copies of X, so that conditional expectation $E^X X' = EX = 0$. Then $Eg(X) = Eg(X - E^X X') = Eg(E^X\{X - X'\})$. Since by Jensen's inequality, see Theorem 1.4.1 (iv) we have $Eg(E^X\{X - X'\}) \leq Eg(X - X')$, therefore $Eg(X) \leq Eg(X - X') = Eg(\widetilde{X}) < \infty$. To end the proof, consider three convex functions $g(x) = |x|^p, g(x) = \exp(\lambda x)$ and $g(x) = \exp(\lambda x^2)$.

1.7   Uniform integrability

Recall that a sequence $\{X_n\}_{n \geq 1}$ is uniformly integrable[4], if

$$\lim_{t \to \infty} \sup_{n \geq 1} \int_{\{|X_n| > t|\}} |X_n| \, dP = 0.$$

Uniform integrability is often used in conjunction with weak convergence to verify the convergence of moments. Namely, if X_n is uniformly integrable and converges in distribution to Y, then Y is integrable and

$$EY = \lim_{n \to \infty} EX_n. \tag{1.18}$$

The following result will be used in the proof of the Central Limit Theorem in Section 7.3.

Proposition 1.7.1 *If $X_1, X_2, \ldots$ are centered i. i. d. random variables with finite second moments and $S_n = \sum_{j=1}^n X_j$ then $\{\frac{1}{n}S_n^2\}_{n \geq 1}$ is uniformly integrable.*

The following lemma is a special case of the celebrated Khinchin inequality.

Lemma 1.7.2 *If ϵ_j are ± 1 valued symmetric independent r. v., then for all real numbers a_j*

$$E\left(\sum_{j=1}^n a_j \epsilon_j\right)^4 \leq 3 \left(\sum_{j=1}^n a_j^2\right)^2 \tag{1.19}$$

---

[4]The contents of this section will be used only in an application part of Section 7.3.

Proof. By independence and symmetry we have

$$E\left(\sum_{j=1}^{n} a_j\epsilon_j\right)^4 = \sum_{j=1}^{n} a_j^4 + 6\sum_{i\neq j} a_i^2 a_j^2$$

which is less than $3\left(\sum_{j=1}^{n} a_j^4 + 2\sum_{i\neq j} a_i^2 a_j^2\right)$. $\square$

The next lemma gives the Marcinkiewicz-Zygmund inequality in the special case needed below.

Lemma 1.7.3 *If X_k are i. i. d. centered with fourth moments, then there is a constant $C < \infty$ such that*

$$ES_n^4 \leq Cn^2 EX_1^4 \tag{1.20}$$

Proof. As in the proof of Theorem 1.6.1 we can estimate the fourth moments of a centered r. v. by the fourth moment of its symmetrization, $ES_n^4 \leq E\widetilde{S}_n^4$.

Let ϵ_j be independent of $\widetilde{X}_k$'s as in Lemma 1.7.2. Then in distribution $\widetilde{S}_n \cong \sum_{j=1}^{n} \epsilon_j \widetilde{X}_j$. Therefore, integrating with respect to the distribution of ϵ_j first, from (1.19) we get

$$ES_n^4 \leq 3E\left(\sum_{j=1}^{n} \widetilde{X}_j^2\right)^2 = 3E\sum_{i,j=1}^{n} \widetilde{X}_i^2 \widetilde{X}_j^2 \leq 3n^2 E\widetilde{X}_1^4.$$

Since $\|X - X'\|_4 \leq 2\|X\|_4$ by triangle inequality (1.7), this ends the proof with $C = 3\cdot 2^4$. $\square$

We shall also need the following inequality.

Lemma 1.7.4 *If $U, V \geq 0$ then*

$$\int_{U+V>2t} (U + V)^2\, dP \leq 4\left(\int_{U>t} U^2\, dP + \int_{V>t} V^2\, dP\right).$$

Proof. By (1.2) applied to $f(x) = x^2 I_{x>2t}$ we have

$$\int_{U+V>2t} (U + V)^2\, dP = \int_{2t}^{\infty} 2x P(U + V > x)\, dx.$$

Since $P(U + V > x) \leq P(U > x/2) + P(V > x/2)$, we get

$$\int_{U+V>2t} (U+V)^2\, dP \leq 4\int_{t}^{\infty} (2yP(U > y) + 2yP(V > y))\, dy = 4\int_{U>t} U^2\, dP + 4\int_{V>t} V^2\, dP.$$

$\square$

Proof of Proposition 1.7.1. We follow Billingsley [10, page 176].

Let $\epsilon > 0$ and choose $M > 0$ such that $\int_{\{|X|>M\}} |X|\, dP < \epsilon$. Split $X_k = X_k' + X_k''$, where $X_k' = X_k I_{\{|X_k|\leq M\}} - E\{X_k I_{\{|X_k|\leq M\}}\}$ and let S', S'' denote the corresponding sums.

Notice that for any $U \geq 0$ we have $U I_{\{|U|>m\}} \leq U^2/m$. Therefore $\frac{1}{n} \int_{|S_n'|>t\sqrt{n}}(S_n')^2 \, dP \leq t^{-2}n^{-2}E(S_n')^4$, which by Lemma 1.7.3 gives

$$\frac{1}{n} \int_{|S_n'|>t\sqrt{n}}(S_n')^2 \, dP \leq CM^4/t^2. \tag{1.21}$$

Now we use orthogonality to estimate the second term:

$$\frac{1}{n} \int_{|S_n''|>t\sqrt{n}}(S_n'')^2 \, dP \leq \frac{1}{n}E(S_n'')^2 \leq E|X_1''|^2 < \epsilon \tag{1.22}$$

To end the proof notice that by Lemma 1.7.4 and inequalities (1.21), (1.22) we have

$$\frac{1}{n} \int_{\{|S_n|>2t\sqrt{n}\}} S_n^2 \, dP \leq \frac{1}{n} \int_{\{|S_n'|+|S_n''|>2t\sqrt{n}\}}(|S_n'| + |S_n''|)^2 \, dP \leq \frac{CM^4}{t^2} + \epsilon.$$

Therefore $\limsup_{t\to\infty} \sup_n \frac{1}{n} \int_{\{|S_n|>2t\sqrt{n}\}} S_n^2 \, dP \leq \epsilon$. Since $\epsilon > 0$ is arbitrary, this ends the proof. $\square$

1.8   The Mellin transform

Definition 1.8.1 [5] *The Mellin transform of a random variable $X \geq 0$ is defined for all complex s such that $EX^{\Re s-1} < \infty$ by the formula $\mathcal{M}(s) = EX^{s-1}$.*

The definition is consistent with the usual definition of the Mellin transform of an integrable function: if X has a probability density function $f(x)$, then the Mellin transform of X is given by $\mathcal{M}(s) = \int_0^\infty x^{s-1} f(x) \, dx$.

Theorem 1.8.1 [6] *If $X \geq 0$ is a random variable such that $EX^{a-1} < \infty$ for some $a \geq 1$, then the Mellin transform $\mathcal{M}(s) = EX^{s-1}$, considered for $s \in \mathbf{C}$ such that $\Re s = a$, determines the distribution of X uniquely.*

Proof. The easiest case is when $a = 1$ and $X > 0$. Then $\mathcal{M}(s)$ is just the characteristic function of $\log(X)$; thus the distribution of $\log(X)$, and hence the distribution of X, is determined uniquely.

In general consider finite non-negative measure μ defined on $(\mathbb{R}_+, \mathcal{B})$ by

$$\mu(A) = \int_{X^{-1}(A)} X^{a-1} \, dP.$$

Then $\mathcal{M}(s)/\mathcal{M}(a)$ is the characteristic function of a random variable $\xi : x \mapsto \log(x)$ defined on the probability space $(\mathbb{R}_+, \mathcal{B}, P')$ with the probability distribution $P'(.) = \mu(.)/\mu(\mathbb{R}_+)$. Thus the distribution of ξ is determined uniquely by $\mathcal{M}(s)$. Since e^ξ has distribution $P'(.)$, μ is determined uniquely by $\mathcal{M}(.)$. It remains to notice that if F is the distribution of our original random variable X, then $dF = x^{1-a}\mu(dx) + \mu(\mathbb{R}_+)\delta_0(dx)$, so $F(.)$ is determined uniquely, too. $\square$

---

[5]The contents of this section has more special character and will only be used in Sections 4.2 and 4.1.
[6]See eg. [152].

Theorem 1.8.2 *If $X \geq 0$ and $EX^a < \infty$ for some $a > 0$, then the Mellin transform of X is analytic in the strip $1 < \Re s < 1 + a$.*

Proof. Since for every s with $0 < \Re s < a$ the modulus of the function $\omega \mapsto X^s \log(X)$ is bounded by an integrable function $C_1 + C_2|X|^a$, therefore EX^s can be differentiated with respect to s under the expectation sign at each point $s, 0 < \Re s < a$. $\square$

1.9 Problems

Problem 1.1 ([64]) *Use Fubini's theorem to show that if XY, X, Y are integrable, then*

$$EXY - EXEY = \int_{-\infty}^{\infty} \int_{-\infty}^{\infty} (P(X \geq t, Y \geq s) - P(X \geq t)P(Y \geq s)) \, dt \, ds.$$

Problem 1.2 *Let $X \geq 0$ be a random variable and suppose that for every $0 < q < 1$ there is $T = T(q)$ such that*

$$P(X > 2t) \leq qP(X > t) \text{ for all } t > T.$$

Show that all the moments of X are finite.

Problem 1.3 *Show that if $X \geq 0$ is a random variable such that*

$$P(X > 2t) \leq (P(X > t))^2 \text{ for all } t > 0,$$

then $E\exp(\lambda|X|) < \infty$ for some $\lambda > 0$.

Problem 1.4 *Show that if $E\exp(\lambda X^2) = C < \infty$ for some $a > 0$, then*

$$E\exp(tX) \leq C \exp(\frac{t^2}{2\lambda})$$

for all real t.

Problem 1.5 *Show that (1.11) implies $E\left\{|X|^{|X|}\right\} < \infty$.*

Problem 1.6 *Prove part (v) of Theorem 1.4.1.*

Problem 1.7 *Prove part (vi) of Theorem 1.4.1.*

Problem 1.8 *Prove part (vii) of Theorem 1.4.1.*

Problem 1.9 *Prove the following conditional version of Chebyshev's inequality: if $\mathcal{F}$ is a σ-field, and $E|X| < \infty$, then*

$$P(|X| > t|\mathcal{F}) \leq E\{|X| \, |\mathcal{F}\}/t$$

almost surely.

Problem 1.10 *Show that if (X, Y) is uniformly distributed on a circle centered at $(0, 0)$, then for every a, b there is a non-random constant $C = C(a, b)$ such that $E\{X | aX + bY\} = C(a, b)(aX + bY)$.*

Problem 1.11 *Show that if (U, V, X) are such that in distribution $(U, X) \cong (V, X)$ then $E\{U | X\} = E\{V | X\}$ almost surely.*

Problem 1.12 *Show that if X, Y are integrable non-degenerate random variables, such that*

$$E\{X | Y\} = aY, \ E\{Y | X\} = bX,$$

then $|ab| \leq 1$.

Problem 1.13 *Suppose that X, Y are square-integrable random variables such that*

$$E\{X | Y\} = Y, \ E\{Y | X\} = 0.$$

Show that $Y = 0$ almost surely[7].

Problem 1.14 *Show that if X, Y are integrable such that $E\{X | Y\} = Y$ and $E\{Y | X\} = X$, then $X = Y$ a. s.*

Problem 1.15 *Prove that if $X \geq 0$, then function $\phi(t) := EX^{it}$, where $t \in \mathbb{R}$, determines the distribution of X uniquely.*

Problem 1.16 *Prove that function $\phi(t) := E\max\{X, t\}$ determines uniquely the distribution of an integrable random variable X in each of the following cases:*

(a) If X is discrete.

(b) If X has continuous density.

Problem 1.17 *Prove that, if $E|X| < \infty$, then function $\phi(t) := E|X - t|$ determines uniquely the distribution of X.*

Problem 1.18 *Let $p > 2$ be fixed. Show that $\exp(-|t|^p)$ is not a characteristic function.*

Problem 1.19 *Let $Q(t, s) = \log \phi(t, s)$, where $\phi(t, s)$ is the joint characteristic function of square-integrable r. v. X, Y.*

(i) Show that $E\{X | Y\} = \rho Y$ implies

$$\left. \frac{\partial}{\partial t} Q(t, s) \right|_{t=0} = \rho \frac{d}{ds} Q(0, s).$$

---

[7]There are, however, non-zero random variables X, Y with this properties, when square-integrability assumption is dropped, see [77].

(ii) Show that $E\{X^2|Y\} = a + bY + cY^2$ implies

$$\left.\frac{\partial^2}{\partial t^2}Q(t,s)\right|_{t=0} + \left.\left(\frac{\partial}{\partial t}Q(t,s)\right)^2\right|_{t=0}$$

$$= -a + ib\frac{d}{ds}Q(0,s) + c\frac{d^2}{ds^2}Q(0,s) + c\left(\frac{d}{ds}Q(0,s)\right)^2.$$

Problem 1.20 (see eg. [76]) *Suppose $a \in \mathbb{R}$ is the median of X.*

(i) Show that the following symmetrization inequality

$$P(|X| \geq t) \leq 2P(|\widetilde{X}| \geq t - |a|)$$

holds for all $t > |a|$.

(ii) Use this inequality to prove Theorem 1.6.1 in the general case.

Problem 1.21 *Suppose (X_n, Y_n) converge to (X, Y) in distribution and $\{X_n\}$, $\{Y_n\}$ are uniformly integrable. If $E(X_n|Y_n) = \rho Y_n$ for all n, show that $E(X|Y) = \rho Y$.*

Problem 1.22 *Prove (1.18).*

Chapter 2

Normal distributions

In this chapter we use linear algebra and characteristic functions to analyze the multivariate normal random variables. More information and other approaches can be found, eg. in [113, 120, 145]. In Section 2.5 we give criteria for normality which will be used often in proofs in subsequent chapters.

2.1 Univariate normal distributions

The usual definition of the standard normal variable Z specifies its density $f(x) = \frac{1}{\sqrt{2\pi}}e^{-\frac{x^2}{2}}$. In general, the so called $N(m, \sigma)$ density is given by

$$f(x) = \frac{1}{\sqrt{2\pi}\sigma}e^{-\frac{(x-m)^2}{2\sigma^2}}.$$

By completing the square one can check that the characteristic function $\phi(t) = Ee^{itZ} = \int_{-\infty}^{\infty} e^{itx}f(x)\,dx$ of the standard normal r. v. Z is given by

$$\phi(t) = e^{-\frac{t^2}{2}},$$

see Problem 2.1.

In multivariate case it is more convenient to use characteristic functions directly. Besides, characteristic functions are our main technical tool and it doesn't hurt to start using them as soon as possible. We shall therefore begin with the following definition.

Definition 2.1.1 *A real valued random variable X has the normal $N(m, \sigma)$ distribution if its characteristic function has the form*

$$\phi(t) = \exp(itm - \frac{1}{2}\sigma^2 t^2),$$

where m, σ are real numbers.

From Theorem 1.5.1 it is easily to check by direct differentiation that $m = EX$ and $\sigma^2 = Var(X)$. Using (1.15) it is easy to see that every univariate normal X can be written as

$$X = \sigma Z + m, \tag{2.1}$$

where Z is the standard $N(0,1)$ random variable with the characteristic function $e^{-\frac{t^2}{2}}$.
The following properties of standard normal distribution $N(0,1)$ are self-evident:

1. The characteristic function $e^{-\frac{t^2}{2}}$ has analytic extension $e^{-\frac{z^2}{2}}$ to all complex $z \in \mathbf{C}$. Moreover, $e^{-\frac{z^2}{2}} \neq 0$.

2. Standard normal random variable Z has finite exponential moments $E \exp(\lambda|Z|) < \infty$ for all λ; moreover, $E \exp(\lambda Z^2) < \infty$ for all $\lambda < \frac{1}{2}$ (compare Problem 1.3).

Relation (2.1) translates the above properties to the general $N(m,\sigma)$ distributions. Namely, if X is normal, then its characteristic function has non-vanishing analytic extension to $\mathbf{C}$ and

$$E \exp(\lambda X^2) < \infty$$

for some $\lambda > 0$.

For future reference we state the following simple but useful observation. Computing EX^k for $k = 0, 1, 2$ from Theorem 1.5.1 we immediately get.

Proposition 2.1.1 *A characteristic function which can be expressed in the form $\phi(t) = \exp(at^2 + bt + c)$ for some complex constants a, b, c, corresponds to the normal random variable, ie. $a \in \mathbb{R}$ and $a < 0, b \in i\mathbb{R}$ is imaginary and $c = 0$.*

2.2　Multivariate normal distributions

We follow the usual linear algebra notation. Vectors are denoted by small bold letters $\mathbf{x}, \mathbf{v}, \mathbf{t}$, matrices by capital bold initial letters $\mathbf{A}, \mathbf{B}, \mathbf{C}$ and vector-valued random variables by capital boldface $\mathbf{X}, \mathbf{Y}, \mathbf{Z}$; by the dot we denote the usual dot product in $\mathbb{R}^d$, ie. $\mathbf{x} \cdot \mathbf{y} := \sum_{j=1}^{d} x_j y_j$; $\|\mathbf{x}\| = (\mathbf{x} \cdot \mathbf{x})^{1/2}$ denotes the usual Euclidean norm. For typographical convenience we sometimes write $(a_1, \ldots, a_k)$ for the vector $\begin{bmatrix} a_1 \\ \vdots \\ a_k \end{bmatrix}$. By $\mathbf{A}^T$ we denote the transpose of a matrix $\mathbf{A}$.

Below we shall also consider another scalar product $\langle \cdot, \cdot \rangle$ associated with the normal distribution; the corresponding semi-norm will be denoted by the triple bar $\|\!|\cdot|\!\|$.

Definition 2.2.1 *An $\mathbb{R}^d$-valued random variable $\mathbf{Z}$ is multivariate normal, or Gaussian (we shall use both terms interchangeably; the second term will be preferred in abstract situations) if for every $\mathbf{t} \in \mathbb{R}^d$ the real valued random variable $\mathbf{t} \cdot \mathbf{Z}$ is normal.*

Clearly the distribution of univariate $\mathbf{t} \cdot \mathbf{Z}$ is determined uniquely by its mean $m = m_{\mathbf{t}}$ and its standard deviation $\sigma = \sigma_{\mathbf{t}}$. It is easy to see that $m_{\mathbf{t}} = \mathbf{t} \cdot \mathbf{m}$, where $\mathbf{m} = E\mathbf{Z}$. Indeed, by linearity of the expected value $m_{\mathbf{t}} = E\mathbf{t} \cdot \mathbf{Z} = \mathbf{t} \cdot E\mathbf{Z}$. Evaluating the characteristic function $\phi(s)$ of the real-valued random variable $\mathbf{t} \cdot \mathbf{Z}$ at $s = 1$ we see that the characteristic function of $\mathbf{Z}$ can be written as

$$\phi(\mathbf{t}) = \exp(i\mathbf{t} \cdot \mathbf{m} - \frac{\sigma_{\mathbf{t}}^2}{2}).$$

In order to rewrite this formula in a more useful form, consider the function $B(\mathbf{x}, \mathbf{y})$ of two arguments $\mathbf{x}, \mathbf{y} \in \mathbb{R}^d$ defined by

$$B(\mathbf{x}, \mathbf{y}) = E\{(\mathbf{x} \cdot \mathbf{Z})(\mathbf{y} \cdot \mathbf{Z})\} - (\mathbf{x} \cdot \mathbf{m}_x)(\mathbf{y} \cdot \mathbf{m}_y).$$

That is, $B(\mathbf{x}, \mathbf{y})$ is the covariance of two real-valued (and jointly Gaussian) random variables $\mathbf{x} \cdot \mathbf{Z}$ and $\mathbf{y} \cdot \mathbf{Z}$.

The following observations are easy to check.

- $B(\cdot, \cdot)$ is symmetric, ie. $B(\mathbf{x}, \mathbf{y}) = B(\mathbf{y}, \mathbf{x})$ for all $\mathbf{x}, \mathbf{y}$;

- $B(\cdot, \cdot)$ is a bilinear function, ie. $B(\cdot, \mathbf{y})$ is linear for every fixed $\mathbf{y}$ and $B(\mathbf{x}, \cdot)$ is linear for very fixed $\mathbf{x}$;

- $B(\cdot, \cdot)$ is positive definite, ie. $B(\mathbf{x}, \mathbf{x}) \geq 0$ for all $\mathbf{x}$.

We shall need the following well known linear algebra fact (the proofs are explained below; explicit reference is, eg. [130, Section 6]).

Lemma 2.2.1 *Each bilinear form B has the dot product representation*

$$B(\mathbf{x}, \mathbf{y}) = \mathbf{C}\mathbf{x} \cdot \mathbf{y},$$

where $\mathbf{C}$ is a linear mapping, represented by a $d \times d$ matrix $\mathbf{C} = [c_{i,j}]$. Furthermore, if $B(\cdot, \cdot)$ is symmetric then $\mathbf{C}$ is symmetric, ie. we have $\mathbf{C} = \mathbf{C}^T$.

Indeed, expand $\mathbf{x}$ and $\mathbf{y}$ with respect to the standard orthogonal basis $\mathbf{e}_1, \ldots, \mathbf{e}_d$. By bilinearity we have $B(\mathbf{x}, \mathbf{y}) = \sum_{i,j} x_i y_j B(\mathbf{e}_i, \mathbf{e}_j)$, which gives the dot product representation with $c_{i,j} = B(\mathbf{e}_i, \mathbf{e}_j)$. Clearly, for symmetric $B(\cdot, \cdot)$ we get $c_{i,j} = c_{j,i}$; hence $\mathbf{C}$ is symmetric.

Lemma 2.2.2 *If in addition $B(\cdot, \cdot)$ is positive definite then*

$$\mathbf{C} = \mathbf{A} \times \mathbf{A}^T \tag{2.2}$$

for a $d \times d$ matrix $\mathbf{A}$. Moreover, $\mathbf{A}$ can be chosen to be symmetric.

The easiest way to see the last fact is to diagonalize $\mathbf{C}$ (this is always possible, as $\mathbf{C}$ is symmetric). The eigenvalues of $\mathbf{C}$ are real and, since $B(\cdot, \cdot)$ is positive definite, they are non-negative. If Λ denotes a (diagonal) matrix (consisting of eigenvalues of $\mathbf{C}$) in the diagonal representation $\mathbf{C} = \mathbf{U}\Lambda\mathbf{U}^T$ and Δ is the diagonal matrix formed by the square roots of the eigenvalues, then $\mathbf{A} = \mathbf{U}\Delta\mathbf{U}^T$. Moreover, this construction gives symmetric $\mathbf{A} = \mathbf{A}^T$. In general, there is no unique choice of $\mathbf{A}$ and we shall sometimes find it more convenient to use non-symmetric $\mathbf{A}$, see Example 2.2.2 below.

The linear algebra results imply that the characteristic function corresponding to a normal distribution on $\mathbb{R}^d$ can be written in the form

$$\phi(\mathbf{t}) = \exp(i\mathbf{t} \cdot \mathbf{m} - \frac{1}{2}\mathbf{C}\mathbf{t} \cdot \mathbf{t}). \tag{2.3}$$

Theorem 1.5.2 identifies $\mathbf{m} \in \mathbb{R}^d$ as the mean of the normal random variable $\mathbf{Z} = (Z_1, \ldots, Z_d)$; similarly, double differentiation $\phi(\mathbf{t})$ at $\mathbf{t} = 0$ shows that $\mathbf{C} = [c_{i,j}]$ is given by $c_{i,j} = Cov(Z_i, Z_j)$. This establishes the following.

Theorem 2.2.3 *The characteristic function corresponding to a normal random variable* $\mathbf{Z} = (Z_1, \ldots, Z_d)$ *is given by (2.3), where* $\mathbf{m} = E\mathbf{Z}$ *and* $\mathbf{C} = [c_{i,j}], c_{i,j} = Cov(Z_i, Z_j)$, *is the covariance matrix.*

From (2.2) and (2.3) we get also

$$\phi(\mathbf{t}) = \exp(i\mathbf{t} \cdot \mathbf{m} - \frac{1}{2}(\mathbf{At}) \cdot (\mathbf{At})). \tag{2.4}$$

In the centered case it is perhaps more intuitive to write $B(\mathbf{x}, \mathbf{y}) = \langle \mathbf{x}, \mathbf{y} \rangle$; this bilinear product might (in degenerate cases) turn out to be 0 on some non-zero vectors. In this notation (2.4) can be written as

$$E\exp(i\mathbf{t} \cdot \mathbf{Z}) = \exp -\frac{1}{2}\langle \mathbf{t}, \mathbf{t} \rangle. \tag{2.5}$$

From the above discussion, we have the following multivariate generalization of (2.1).

Theorem 2.2.4 *Each d-dimensional normal random variable* $\mathbf{Z}$ *has the same distribution as* $\mathbf{m} + \mathbf{A}\vec{\gamma}$, *where* $\mathbf{m} \in \mathbb{R}^d$ *is deterministic,* $\mathbf{A}$ *is a (symmetric)* $d \times d$ *matrix and* $\vec{\gamma} = (\gamma_1, \ldots, \gamma_d)$ *is a random vector such that the components* $\gamma_1, \ldots, \gamma_d$ *are independent* $N(0,1)$ *random variables.*

Proof. Clearly, $E\exp(i\mathbf{t} \cdot (\mathbf{m} + \mathbf{A}\vec{\gamma})) = \exp(i\mathbf{t} \cdot \mathbf{m})E\exp(i\mathbf{t} \cdot (\mathbf{A}\vec{\gamma}))$. Since the characteristic function of $\vec{\gamma}$ is $E\exp(i\mathbf{x} \cdot \vec{\gamma}) = \exp -\frac{1}{2}\|\mathbf{x}\|^2$ and $\mathbf{t} \cdot (\mathbf{A}\vec{\gamma}) = (\mathbf{A}^T\mathbf{t}) \cdot \vec{\gamma}$, therefore we get $E\exp(i\mathbf{t} \cdot (\mathbf{m} + \mathbf{A}\vec{\gamma})) = \exp i\mathbf{t} \cdot \mathbf{m} \exp -\frac{1}{2}\|\mathbf{A}^T\mathbf{t}\|^2$, which is another form of (2.4). $\square$

Theorem 2.2.4 can be actually interpreted as the almost sure representation. However, if $\mathbf{A}$ is not of full rank, the number of independent $N(0,1)$ r. v. can be reduced. In addition, the representation $\mathbf{Z} \cong \mathbf{m} + \mathbf{A}\vec{\gamma}$ from Theorem 2.2.4 is not unique if the symmetry condition is dropped. Theorem 2.2.5 gives the same representation with non-symmetric $\mathbf{A} = [\mathbf{e}_1, \ldots, \mathbf{e}_k]$. The argument given below has more geometric flavor. Infinite dimensional generalizations are also known, see (8.4) and the comment preceding Lemma 8.1.1.

Theorem 2.2.5 *Each d-dimensional normal random variable* $\mathbf{Z}$ *can be written as*

$$\mathbf{Z} = \mathbf{m} + \sum_{j=1}^{k} \gamma_j \mathbf{e}_j, \tag{2.6}$$

where $k \leq d, \mathbf{m} \in \mathbb{R}^d, \mathbf{e}_1, \mathbf{e}_2, \ldots, \mathbf{e}_k$ *are deterministic linearly independent vectors in* $\mathbb{R}^d$ *and* $\gamma_1, \ldots, \gamma_k$ *are independent identically distributed normal* $N(0,1)$ *random variables.*

Proof. Without loss of generality we may assume $E\mathbf{Z} = 0$ and establish the representation with $\mathbf{m} = 0$.

Let $\mathbb{H}$ denote the linear span of the columns of $\mathbf{A}$ in $\mathbb{R}^d$, where $\mathbf{A}$ is the matrix from (2.4). From Theorem 2.2.4 it follows that with probability one $\mathbf{Z} \in \mathbb{H}$. Consider now $\mathbb{H}$ as a Hilbert space with a scalar product $\langle \mathbf{x}, \mathbf{y} \rangle$, given by $\langle \mathbf{x}, \mathbf{y} \rangle = (\mathbf{Ax}) \cdot (\mathbf{Ay})$. Since the null space of $\mathbf{A}$ and the column space of $\mathbf{A}$ have only zero vector in common, this scalar product is non-degenerate, ie. $\langle \mathbf{x}, \mathbf{x} \rangle \neq 0$ for $\mathbb{H} \ni \mathbf{x} \neq 0$.

Let $\mathbf{e}_1, \mathbf{e}_2, \ldots, \mathbf{e}_k$ be the orthonormal (with respect to $\langle \cdot, \cdot \rangle$) basis of $\mathbb{H}$, where $k = \dim \mathbb{H}$. By Theorem 2.2.4 $\mathbf{Z}$ is $\mathbb{H}$-valued. Therefore with probability one we can write $\mathbf{Z} = \sum_{j=1}^{k} \gamma_j \mathbf{e}_j$, where $\gamma_j = \langle \mathbf{e}_j, \mathbf{Z} \rangle$ are random coefficients in the orthogonal expansion. It remains to verify that $\gamma_1, \ldots, \gamma_k$ are i. i. d. normal $N(0,1)$ r. v. With this in mind, we use (2.4) to compute their joint characteristic function:

$$E \exp(i \sum_{j=1}^{k} t_j \gamma_j) = E \exp(i \sum_{j=1}^{k} t_j \langle \mathbf{e}_j, \mathbf{Z} \rangle) = E \exp(i \langle \sum_{j=1}^{k} t_j \mathbf{e}_j, \mathbf{Z} \rangle).$$

By (2.5)

$$E \exp(i \langle \sum_{j=1}^{k} t_j \mathbf{e}_j, \mathbf{Z} \rangle) = \exp(-\frac{1}{2} \langle \sum_{j=1}^{k} t_j \mathbf{e}_j, \sum_{j=1}^{k} t_j \mathbf{e}_j \rangle) = \exp(-\frac{1}{2} \sum_{j=1}^{k} t_j^2).$$

The last equality is a consequence of orthonormality of vectors $\mathbf{e}_1, \mathbf{e}_2, \ldots, \mathbf{e}_k$ with respect to the scalar product $\langle \cdot, \cdot \rangle$. $\square$

The next theorem lists two important properties of the normal distribution that can be easily verified by writing the joint characteristic function. The second property is a consequence of the polarization identity

$$\|t + s\|^2 + \|t - s\|^2 = \|t\|^2 + \|s\|^2,$$

where

$$\|\mathbf{x}\|^2 := \langle \mathbf{x}, \mathbf{x} \rangle := \|\mathbf{A}\mathbf{x}\|^2; \tag{2.7}$$

the proof is left as an exercise.

Theorem 2.2.6 *If* $\mathbf{X}, \mathbf{Y}$ *are independent with the same centered normal distribution, then*

a) $\frac{\mathbf{X}+\mathbf{Y}}{\sqrt{2}}$ *has the same distribution as* $\mathbf{X}$;
b) $\mathbf{X} + \mathbf{Y}$ *and* $\mathbf{X} - \mathbf{Y}$ *are independent.*

Now we consider the multivariate normal density. The density of $\vec{\gamma}$ in Theorem 2.2.4 is the product of the one-dimensional standard normal densities, ie.

$$f_{\vec{\gamma}}(\mathbf{x}) = (2\pi)^{-d/2} \exp(-\frac{1}{2}\|\mathbf{x}\|^2).$$

Suppose that $\det \mathbf{C} \neq 0$, which ensures that $\mathbf{A}$ is nonsingular. By the change of variable formula, from Theorem 2.2.4 we get the following expression for the multivariate normal density.

Theorem 2.2.7 *If* $\mathbf{Z}$ *is centered normal with the nonsingular covariance matrix* $\mathbf{C}$*, then the density of* $\mathbf{Z}$ *is given by*

$$f_{\mathbf{Z}}(\mathbf{x}) = (2\pi)^{-d/2} (\det \mathbf{A})^{-1} \exp(-\frac{1}{2}\|\mathbf{A}^{-1}\mathbf{x}\|^2),$$

or

$$f_{\mathbf{Z}}(\mathbf{x}) = (2\pi)^{-d/2} (\det \mathbf{C})^{-1/2} \exp(-\frac{1}{2}\mathbf{C}^{-1}\mathbf{x} \cdot \mathbf{x}),$$

where matrices $\mathbf{A}$ *and* $\mathbf{C}$ *are related by (2.2).*

In the nonsingular case this immediately implies strong integrability.

Theorem 2.2.8 *If* $\mathbf{Z}$ *is normal, then there is* $\epsilon > 0$ *such that*

$$E\exp(\epsilon\|\mathbf{Z}\|^2) < \infty.$$

Remark: Theorem 2.2.8 holds true also in the singular case and for Gaussian random variables with values in infinite dimensional spaces; for the proof based on Theorem 2.2.6, see Theorem 5.4.2 below.

The Hilbert space $\mathbb{H}$ introduced in the proof of Theorem 2.2.5 is called the *Reproducing Kernel Hilbert Space* (RKHS) of a normal distribution, cf. [5, 90]. It can be defined also in more general settings. Suppose we want to consider jointly two independent normal r. v. $\mathbf{X}$ and $\mathbf{Y}$, taking values in $\mathbb{R}^{d_1}$ and $\mathbb{R}^{d_2}$ respectively, with corresponding reproducing kernel Hilbert spaces $\mathbb{H}_1, \mathbb{H}_2$ and the corresponding dot products $\langle\cdot,\cdot\rangle_1$ and $\langle\cdot,\cdot\rangle_2$. Then the $\mathbb{R}^{d_1+d_2}$-valued random variable $(\mathbf{X},\mathbf{Y})$ has the orthogonal sum $\mathbb{H}_1 \oplus \mathbb{H}_2$ as the Reproducing Kernel Hilbert Space.

This method shows further geometric aspects of jointly normal random variables. Suppose an $\mathbb{R}^{d_1+d_2}$-valued random variable $(\mathbf{X},\mathbf{Y})$ is (jointly) normal and has $\mathbb{H}$ as the reproducing kernel Hilbert space (with the scalar product $\langle\cdot,\cdot\rangle$). Recall that $\mathbb{H} = \left\{\mathbf{A}\begin{bmatrix}\mathbf{x}\\\mathbf{y}\end{bmatrix} : \begin{bmatrix}\mathbf{x}\\\mathbf{y}\end{bmatrix} \in \mathbb{R}^{d_1+d_2}\right\}$. Let $\mathbb{H}_Y$ be the subspace of $\mathbb{H}$ spanned by the vectors $\left\{\begin{bmatrix}\mathbf{0}\\\mathbf{y}\end{bmatrix} : \mathbf{y} \in \mathbb{R}^{d_2}\right\}$; similarly let $\mathbb{H}_X$ be the subspace of $\mathbb{H}$ spanned by the vectors $\begin{bmatrix}\mathbf{x}\\\mathbf{0}\end{bmatrix}$. Let $\mathbf{P}$ be (the matrix of) the linear transformation $\mathbb{H}_X \to \mathbb{H}_Y$ obtained from the $\langle\cdot,\cdot\rangle$-orthogonal projection $\mathbb{H} \to \mathbb{H}_X$ by narrowing its domain to $\mathbb{H}_X$. Denote $\mathbf{Q} = \mathbf{P}^T$; $\mathbf{Q}$ represents the orthogonal projection in the *dual* norm defined in Section 2.6 below.

Theorem 2.2.9 *If* $(\mathbf{X},\mathbf{Y})$ *has jointly normal distribution on* $\mathbb{R}^{d_1+d_2}$, *then random vectors* $\mathbf{Y} - \mathbf{Q}\mathbf{Y}$ *and* $\mathbf{X}$ *are stochastically independent.*

Proof. The joint characteristic function of $\mathbf{X} - \mathbf{Q}\mathbf{Y}$ and $\mathbf{Y}$ factors as follows:

$$\phi(\mathbf{t},\mathbf{s}) = E\exp(i\mathbf{t} \cdot (\mathbf{X} - \mathbf{Q}\mathbf{Y}) + i\mathbf{s} \cdot \mathbf{Y})$$

$$= E\exp(i\mathbf{t} \cdot \mathbf{X} - \mathbf{P}\mathbf{t} \cdot \mathbf{Y} + i\mathbf{s} \cdot \mathbf{Y})$$

$$= \exp(-\tfrac{1}{2}\left\|\begin{bmatrix}\mathbf{t}\\\mathbf{s} - \mathbf{P}\mathbf{t}\end{bmatrix}\right\|^2) = \exp(-\tfrac{1}{2}\left\|\begin{bmatrix}\mathbf{t}\\-\mathbf{P}\mathbf{t}\end{bmatrix}\right\|^2)\exp(-\tfrac{1}{2}\left\|\begin{bmatrix}\mathbf{0}\\\mathbf{s}\end{bmatrix}\right\|^2).$$

The last identity holds because by our choice of $\mathbf{P}$, vectors $\begin{bmatrix}\mathbf{0}\\\mathbf{s}\end{bmatrix}$ and $\begin{bmatrix}\mathbf{t}\\-\mathbf{P}\mathbf{t}\end{bmatrix}$ are orthogonal with respect to scalar product $\langle\cdot,\cdot\rangle$. $\square$

In particular, since $E\{\mathbf{X}|\mathbf{Y}\} = E\{\mathbf{X} - \mathbf{Q}\mathbf{Y}|\mathbf{Y}\} + \mathbf{Q}\mathbf{Y}$, we get

Corollary 2.2.10 *If both* $\mathbf{X}$ *and* $\mathbf{Y}$ *have mean zero, then*

$$E\{\mathbf{X}|\mathbf{Y}\} = \mathbf{Q}\mathbf{Y}.$$

For general multivariate normal random variables $\mathbf{X}$ and $\mathbf{Y}$ applying the above to centered normal random variables $\mathbf{X} - \mathbf{m_X}$ and $\mathbf{Y} - \mathbf{m_Y}$ respectively, we get

$$E\{\mathbf{X}|\mathbf{Y}\} = \mathbf{a} + \mathbf{QY}; \tag{2.8}$$

vector $\mathbf{a} = \mathbf{m_X} - \mathbf{Q}\mathbf{m_Y}$ and matrix $\mathbf{Q}$ are determined by the expected values $\mathbf{m_X}, \mathbf{m_Y}$ and by the (joint) covariance matrix $\mathbf{C}$ (uniquely if the covariance $\mathbf{C_Y}$ of $\mathbf{Y}$ is non-singular). To find $\mathbf{Q}$, multiply (2.8) (as a column vector) from the right by $(\mathbf{Y} - E\mathbf{Y})^T$ and take the expected value. By Theorem 1.4.1(i) we get $\mathbf{Q} = \mathbf{R} \times \mathbf{C_Y^{-1}}$, where we have written $\mathbf{C}$ as the (suitable) block matrix $\mathbf{C} = \begin{bmatrix} \mathbf{C_X} & \mathbf{R} \\ \mathbf{R}^T & \mathbf{C_Y} \end{bmatrix}$. An alternative proof of (2.8) (and of Corollary 2.2.10) is to use the converse to Theorem 1.5.3.

Equality (2.8) is usually referred to as linearity of regression. For the bivariate normal distribution it takes the form $E\{X|Y\} = \alpha + \beta Y$ and it can be established by direct integration; for more than two variables computations become more difficult and the characteristic functions are quite handy.

Corollary 2.2.11 *Suppose $(\mathbf{X}, \mathbf{Y})$ has a (joint) normal distribution on $\mathbb{R}^{d_1 + d_2}$ and $\mathbb{H}_X, \mathbb{H}_Y$ are $\langle \cdot, \cdot, \cdot \rangle$-orthogonal, ie. every component of $\mathbf{X}$ is uncorrelated with all components of $\mathbf{Y}$. Then $\mathbf{X}, \mathbf{Y}$ are independent.*

Indeed, in this case $\mathbf{Q}$ is the zero matrix; the conclusion follows from Theorem 2.2.9.

Example 2.2.1 *In this example we consider a pair of (jointly) normal random variables X_1, X_2. For simplicity of notation we suppose $EX_1 = 0, EX_2 = 0$. Let $Var(X_1) = \sigma_1^2, Var(X_2) = \sigma_2^2$ and denote $corr(X_1, X_2) = \rho$. Then $\mathbf{C} = \begin{bmatrix} \sigma_1^2 & \rho \\ \rho & \sigma_2^2 \end{bmatrix}$ and the joint characteristic function is*

$$\phi(t_1, t_2) = \exp(-\frac{1}{2}t_1^2\sigma_1^2 - \frac{1}{2}t_2^2\sigma_2^2 - t_1 t_2 \rho).$$

If $\sigma_1\sigma_2 \neq 0$ we can normalize the variables and consider the pair $Y_1 = X_1/\sigma_1$ and $Y_2 = X_2/\sigma_2$. The covariance matrix of the last pair is $\mathbf{C_Y} = \begin{bmatrix} 1 & \rho \\ \rho & 1 \end{bmatrix}$; the corresponding scalar product is given by

$$\left\langle \begin{bmatrix} x_1 \\ x_2 \end{bmatrix}, \begin{bmatrix} y_1 \\ y_2 \end{bmatrix} \right\rangle = x_1 y_1 + x_2 y_2 + \rho x_1 y_2 + \rho x_2 y_1$$

and the corresponding RKHS norm is $\left\| \begin{bmatrix} x_1 \\ x_2 \end{bmatrix} \right\| = (x_1^2 + x_2^2 + 2\rho x_1 x_2)^{1/2}$. Notice that when $\rho = \pm 1$ the RKHS norm is degenerate and equals $|x_1 \pm x_2|$.

Denoting $\rho = \sin 2\theta$, it is easy to check that $\mathbf{A_Y} = \begin{bmatrix} \cos\theta & \sin\theta \\ \sin\theta & \cos\theta \end{bmatrix}$ and its inverse $\mathbf{A_Y^{-1}} = \frac{1}{\cos 2\theta} \begin{bmatrix} \cos\theta & -\sin\theta \\ -\sin\theta & \cos\theta \end{bmatrix}$ exists if $\theta \neq \pm\pi/4$, ie. when $\rho^2 \neq 1$. This implies that the joint density of Y_1 and Y_2 is given by

$$f(x, y) = \frac{1}{2\pi\cos 2\theta}\exp(-\frac{1}{2\cos^2 2\theta}(x^2 + y^2 - 2xy\sin 2\theta)). \tag{2.9}$$

We can easily verify that in this case Theorem 2.2.5 gives

$$Y_1 = \gamma_1 \cos\theta + \gamma_2 \sin\theta,$$

$$Y_2 = \gamma_1 \sin\theta + \gamma_2 \cos\theta$$

for some i.i.d normal $N(0,1)$ r. v. γ_1, γ_2. One way to see this, is to compare the variances and the covariances of both sides. Another representation $Y_1 = \gamma_1$, $Y_2 = \rho\gamma_1 + \sqrt{1-\rho^2}\gamma_2$ illustrates non-uniqueness and makes Theorem 2.2.9 obvious in bivariate case.

Returning back to our original random variables X_1, X_2, we have $X_1 = \gamma_1\sigma_1\cos\theta + \gamma_2\sigma_1\sin\theta$ and $X_2 = \gamma_1\sigma_2\sin\theta + \gamma_2\sigma_2\cos\theta$; this representation holds true also in the degenerate case.

To illustrate previous theorems, notice that Corollary 2.2.11 in the bivariate case follows immediately from (2.9). Theorem 2.2.9 says in this case that $Y_1 - \rho Y_2$ and Y_2 are independent; this can also be easily checked either by using density (2.9) directly, or (easier) by verifying that $Y_1 - \rho Y_2$ and Y_2 are uncorrelated.

Example 2.2.2 *In this example we analyze a discrete time Gaussian random walk $\{X_k\}_{0\leq k\leq T}$. Let $\xi_1, \xi_2, \ldots$ be i. i. d. $N(0,1)$. We are interested in explicit formulas for the characteristic function and for the density of the $\mathbb{R}^T$-valued random variable $\mathbf{X} = (X_1, X_2, \ldots, X_T)$, where*

$$X_k = \sum_{j=1}^{k} \xi_j \tag{2.10}$$

are partial sums.

Clearly, $\mathbf{m} = 0$. Comparing (2.10) with (2.6) we observe that

$$\mathbf{A} = \begin{bmatrix} 1 & 0 & \ldots & 0 \\ 1 & 1 & \ldots & 0 \\ \vdots & & \ddots & \vdots \\ 1 & 1 & \ldots & 1 \end{bmatrix}.$$

Therefore from (2.4) we get

$$\phi(\mathbf{t}) = \exp{-\frac{1}{2}(t_1^2 + (t_1 + t_2)^2 + \ldots + (t_1 + t_2 + \ldots + t_T)^2)}.$$

To find the formula for joint density, notice that $\mathbf{A}$ is the matrix representation of the linear operator, which to a given sequence of numbers $(x_1, x_2, \ldots, x_T)$ assigns the sequence of its partial sums $(x_1, x_1 + x_2, \ldots, x_1 + x_2 + \ldots + x_T)$. Therefore, its inverse is the finite difference operator $\Delta : (x_1, x_2, \ldots, x_T) \mapsto (x_1, x_2 - x_1, \ldots, x_T - x_{T-1})$. This implies

$$\mathbf{A}^{-1} = \begin{bmatrix} 1 & 0 & 0 & \ldots & \ldots & 0 \\ -1 & 1 & 0 & \ldots & \ldots & 0 \\ 0 & -1 & 1 & \ldots & \ldots & 0 \\ 0 & 0 & -1 & \ldots & \ldots & 0 \\ \vdots & \ddots & & & \ddots & \vdots \\ 0 & \ldots & 0 & \ldots & -1 & 1 \end{bmatrix}.$$

Since $\det \mathbf{A} = 1$, *we get*

$$f(\mathbf{x}) = (2\pi)^{-n/2} \exp -\frac{1}{2}(x_1^2 + (x_2 - x_1)^2 + \ldots + (x_T - x_{T-1})^2). \qquad (2.11)$$

Interpreting $\mathbf{X}$ *as the discrete time process* $X_1, X_2, \ldots$, *the probability density function for its trajectory* $\mathbf{x}$ *is given by* $f(\mathbf{x}) = C \exp(-\frac{1}{2}\|\Delta\mathbf{x}\|^2)$. *Expression* $\frac{1}{2}\|\Delta\mathbf{x}\|^2$ *can be interpreted as proportional to the kinetic energy of the motion described by the path* $\mathbf{x}$; *assigning probabilities by* $Ce^{-Energy/(kT)}$ *is a well known practice in statistical physics. In continuous time, the derivative plays analogous role, compare Schilder's theorem [34, Theorem 1.3.27].*

2.3 Analytic characteristic functions

The characteristic function $\phi(t)$ of the univariate normal distribution is a well defined differentiable function of complex argument t. That is, ϕ has analytic extension to complex plane $\mathbf{C}$. The theory of functions of complex variable provides a powerful tool; we shall use it to recognize the normal characteristic functions. Deeper theory of analytic characteristic functions and stronger versions of theorems below can be found in monographs [99, 103].

Definition 2.3.1 *We shall say that a characteristic function* $\phi(t)$ *is analytic if it can be extended from the real line* $\mathbb{R}$ *to the function analytic in a domain in complex plane* $\mathbf{C}$.

Because of uniqueness we shall use the same symbol ϕ to denote both.

Clearly, normal distribution has analytic characteristic function. Example 1.5.1 presents a non-analytic characteristic function.

We begin with the probabilistic (moment) condition for the existence of the analytic extension.

Theorem 2.3.1 *If a random variable* X *has finite exponential moment* $E\exp(a|X|) < \infty$, *where* $a > 0$, *then its characteristic function* $\phi(s)$ *is analytic in the strip* $-a < \Im s < a$.

Proof. The analytic extension is given explicitly: $\phi(s) = E\exp(isX)$. It remains only to check that $\phi(s)$ is differentiable in the strip $-a < \Im s < a$. This follows either by differentiation with respect to s under the expectation sign (the latter is allowed, since $E\{|X|\exp(|sX|)\} < \infty$, provided $-a < \Im s < a$), or by writing directly the series expansion: $\phi(s) = \sum_{n=0}^{\infty} i^n E X^n s^n / n!$ (the last equality follows by switching the order of integration and summation, ie. by Fubini's theorem). The series is easily seen to be absolutely convergent for all $-a \leq \Im s \leq a$. $\square$

Corollary 2.3.2 *If* X *is such that* $E\exp(a|X|) < \infty$ *for every real* $a > 0$, *then its characteristic function* $\phi(s)$ *is analytic in* $\mathbf{C}$.

The next result says that normal distribution is determined uniquely by its moments. For more information on the moment problem, the reader is referred to the beautiful book by N. I. Akhiezer [2].

Corollary 2.3.3 *If X is a random variable with finite moments of all orders and such that $EX^k = EZ^k, k = 1, 2, \ldots$, where Z is normal, then X is normal.*

Proof. By the Taylor expansion

$$E\exp(a|X|) = \sum a^k E|X|^k/k! = E\exp(a|Z|) < \infty$$

for all real $a > 0$. Therefore by Corollary 2.3.2 the characteristic function of X is analytic in $\mathbb{C}$ and it is determined uniquely by its Taylor expansion coefficients at 0. However, by Theorem 1.5.1(ii) the coefficients are determined uniquely by the moments of X. Since those are the same as the corresponding moments of the normal r. v. Z, both characteristic functions are equal. $\square$

We shall also need the following refinement of Corollary 2.3.3.

Corollary 2.3.4 *Let $\phi(t)$ be a characteristic function, and suppose there is $\sigma^2 > 0$ and a sequence $\{t_k\}$ convergent to 0 such that $\phi(t_k) = \exp(-\sigma^2 t_k^2)$ and $t_k \neq 0$ for all k. Then $\phi(t) = \exp(-\sigma^2 t^2)$ for every $t \in \mathbb{R}$.*

Proof. The idea of the proof is simply to calculate all the derivatives at 0 of $\phi(t)$ along the sequence $\{t_k\}$. Since the derivatives determine moments uniquely, by Corollary 2.3.3 we shall conclude that $\phi(t) = \exp(-\sigma^2 t^2)$. The only nuisance is to establish that all the moments of the distribution are finite. This fact is established by modifying the usual proof of Theorem 1.5.1(iii). Let Δ_t^2 be a symmetric second order difference operator, ie.

$$\Delta_t^2(g)(y) := \frac{g(y+t) + g(y-t) - 2g(y)}{t^2}.$$

The assumption that $\phi(t)$ is differentiable $2n$ times along the sequence $\{t_k\}$ implies that

$$\sup_k |\Delta_{t(k)}^{2n}(\phi)(0)| = \sup_k |\Delta_{t(k)}^2 \Delta_{t(k)}^2 \ldots \Delta_{t(k)}^2(\phi)(0)| < \infty.$$

Indeed, the assumption says that $\lim_{k \to \infty} \Delta_{t(k)}^{2n}(\phi)(0)$ exists for all n. Therefore to end the proof we need the following result.

Claim 2.3.1 *If $\phi(t)$ is the characteristic function of a random variable $X, t(k) \to 0$ is a given sequence such that $t(k) \neq 0$ for all k and*

$$\sup_k |\Delta_{t(k)}^{2n}(\phi)(0)| < \infty$$

for an integer n, then $EX^{2n} < \infty$.

The proof of the claim rests on the formula which can be verified by elementary calculations:

$$\{\Delta_t^2 \exp(iay)\}(y)\Big|_{y=x} = 4t^{-2} \exp(iax) \sin^2(at/2).$$

This permits to express recurrently the higher order differences, giving

$$\{\Delta_{t(k)}^{2n} \exp(iay)\}(y)\Big|_{y=x} = 4^n t^{-2n} \sin^{2n}(at/2) \exp(iax).$$

Therefore

$$|\Delta^{2n}_{t(k)}(\phi)(0)| = 4^n t(k)^{-2n} E \sin^{2n}(t(k)X/2)$$

$$\geq 4^n t(k)^{-2n} E 1_{|X|\leq 2/|t(k)|} \sin^{2n}(t(k)X/2).$$

The graph of $\sin(x)$ shows that inequality $|\sin(x)| \geq \frac{2}{\pi}|x|$ holds for all $|x| \leq \frac{\pi}{2}$. Therefore

$$|\Delta^{2n}_{t(k)}(\phi)(0)| \geq \left(\frac{2}{\pi}\right)^{2n} E 1_{|X|\leq 2/|t(k)|} X^{2n}.$$

By the monotone convergence theorem

$$EX^{2n} \leq \limsup_{k\to\infty} E 1_{|X|\leq 2/|t(k)|} X^{2n} < \infty,$$

which ends the proof. $\square$

The next result is converse to Theorem 2.3.1.

Theorem 2.3.5 *If the characteristic function $\phi(t)$ of a random variable X has the analytic extension in a neighborhood of 0 in $\mathbb{C}$, and the extension is such that the Taylor expansion series at 0 has convergence radius $R \leq \infty$, then $E\exp(a|X|) < \infty$ for all $0 \leq a < R$.*

Proof. By assumption, $\phi(s)$ has derivatives of all orders. Thus the moments of all orders are finite and

$$m_k = EX^k = (-i)^k \left.\frac{\partial^k}{\partial s^k}\phi(s)\right|_{s=0}, k \geq 1.$$

Taylor's expansion of $\phi(s)$ at $s = 0$ is given by $\phi(s) = \sum_{k=0}^{\infty} i^k m_k s^k / k!$. The series has convergence radius R if and only if $\limsup_{k\to\infty}(m_k/k!)^{1/k} = 1/R$. This implies that for any $0 \leq a < A < R$, there is k_0, such that $m_k \leq A^{-k} k!$ for all $k \geq k_0$. Hence $E\exp(a|X|) = \sum_{k=0}^{\infty} a^k m_k / k! < \infty$, which ends the proof of the theorem. $\square$

Theorems 2.3.1 and 2.3.5 combined together imply the following.

Corollary 2.3.6 *If a characteristic function $\phi(t)$ can be extended analytically to the circle $|s| < a$, then it has analytic extension $\phi(s) = E\exp(isX)$ to the strip $-a < \Im s < a$.*

2.4 Hermite expansions

A normal N(0,1) r. v. Z defines a dot product $\langle f,g \rangle = Ef(Z)g(Z)$, provided that $f(Z)$ and $g(Z)$ are square integrable functions on Ω. In particular, the dot product is well defined for polynomials. One can apply the usual Gram-Schmidt orthogonalization algorithm to functions $1, Z, Z^2, \ldots$. This produces orthogonal polynomials in variable Z known as Hermite polynomials. Those play important role and can be equivalently defined by

$$H_n(x) = (-1)^n \exp(x^2/2)\frac{d^n}{dx^n}\exp(-x^2/2).$$

Hermite polynomials actually form an orthogonal basis of $L_2(Z)$. In particular, every function f such that $f(Z)$ is square integrable can be expanded as $f(x) = \sum_{n=1}^{\infty} f_k H_k(x)$, where $f_k \in \mathbb{R}$ are Fourier coefficients of $f(\cdot)$; the convergence is in $L_2(Z)$, ie. in weighted L_2 norm on the real line, $L_2(\mathbb{R}, e^{-x^2/2}dx)$.

The following is the classical Mehler's formula.

Theorem 2.4.1 *For a bivariate normal r. v. X, Y with $EX = EY = 0$, $EX^2 = EY^2 = 1$, $EXY = \rho$, the joint density $q(x,y)$ of X, Y is given by*

$$q(x,y) = \sum_{k=0}^{\infty} \rho^k/k! H_k(x)H_k(y)q(x)q(y), \tag{2.12}$$

where $q(x) = (2\pi)^{-1/2} \exp(-x^2/2)$ is the marginal density.

Proof. By Fourier's inversion formula we have

$$q(x,y) = \frac{1}{2\pi} \int \int \exp(itx + ity) \exp(-\frac{1}{2}t^2 - \frac{1}{2}s^2) \exp(-\rho ts) \, dt \, ds.$$

Since $(-1)^k t^k s^k \exp(itx + isy) = \frac{\partial^{2k}}{\partial x^k \partial y^k} \exp(itx + isy)$, expanding $e^{-\rho ts}$ into the Taylor series we get

$$q(x,y) = \sum_{k=0}^{\infty} \frac{\rho^k}{k!} \frac{\partial^{2k}}{\partial x^k \partial y^k} q(x)q(y).$$

$\square$

2.5 Cramer and Marcinkiewicz theorems

The next lemma is a direct application of analytic functions theory.

Lemma 2.5.1 *If X is a random variable such that $E\exp(\lambda X^2) < \infty$ for some $\lambda > 0$, and the analytic extension $\phi(z)$ of the characteristic function of X satisfies $\phi(z) \neq 0$ for all $z \in \mathbf{C}$, then X is normal.*

Proof. By the assumption, $f(z) = \log \phi(z)$ is well defined and analytic for all $z \in \mathbf{C}$. Furthermore if $z = x + iy$ is the decomposition of $z \in \mathbf{C}$ into its real and imaginary parts, then $\Re f(z) = \log |\phi(z)| \leq \log(E\exp |yX|)$. Notice that $E\exp(tX) \leq C \exp(\frac{t^2}{2\lambda})$ for all real t, see Problem 1.4. Indeed, since $\lambda X^2 + t^2/\lambda \geq 2tX$, therefore $E\exp(tX) \leq E\exp(\lambda X^2 + t^2/a)/2 = C \exp(\frac{t^2}{2\lambda})$. Those two facts together imply $\Re f(z) \leq const + \frac{y^2}{2a}$. Therefore a variant of the Liouville theorem [144, page 87] implies that $f(z)$ is a quadratic polynomial in variable z, ie. $f(z) = A + Bz + Cz^2$. It is easy to see that the coefficients are $A = 0$, $B = iE\{X\}$, $C = -Var(X)/2$, compare Proposition 2.1.1. $\square$

From Lemma 2.5.1 we obtain quickly the following important theorem, due to H. Cramer [29].

Theorem 2.5.2 *If X_1 and X_2 are independent random variables such that $X_1 + X_2$ has a normal distribution, then each of the variables X_1, X_2 is normal.*

Theorem 2.5.2 is celebrated Cramer's decomposition theorem; for extensions, see [99]. Cramer's theorem complements nicely the Central Limit Theorem in the following sense. While the Central Limit Theorem asserts that the distribution of the sum of i. i. d. random variables with finite variances is close to normal, Cramer's theorem says that it cannot be exactly normal, except when we start with a normal sequence. This resembles *propagation of chaos* phenomenon, where one proves a dynamical system approaches chaotic behavior, but it never reaches it except from initially chaotic configurations. We shall use Theorem 2.5.2 as a technical tool.

Proof of Theorem 2.5.2. Without loss of generality we may assume $EX_1 = EX_2 = 0$. The proof of Theorem 1.6.1 (iii) implies that $E\exp(aX_j^2) < \infty, j = 1, 2$. Therefore, by Theorem 2.3.1, the corresponding characteristic functions $\phi_1(\cdot), \phi_2(\cdot)$ are analytic. By the uniqueness of the analytic extension, $\phi_1(s)\phi_2(s) = \exp(-s^2/2)$ for all $s \in \mathbf{C}$. Thus $\phi_j(z) \neq 0$ for all $z \in \mathbf{C}, j = 1, 2$, and by Lemma 2.5.1 both characteristic functions correspond to normal distributions. $\square$

The next theorem is useful in recognizing the normal distribution from what at first sight seems to be incomplete information about a characteristic function. The result and the proof come from Marcinkiewicz [106], cf. [105].

Theorem 2.5.3 *Let $Q(t)$ be a polynomial, and suppose that a characteristic function ϕ has the representation $\phi(t) = \exp Q(t)$ for all t close enough to 0. Then Q is of degree at most 2 and ϕ corresponds to a normal distribution.*

Proof. First note that formula $\phi(s) = \exp Q(s)$, $s \in \mathbf{C}$, defines the analytic extension of ϕ. Thus, by Corollary 2.3.6, $\phi(s) = E\exp(isX)$, $s \in \mathbf{C}$. By Theorem 2.5.2, it suffices to show that $\phi(s)\phi(-s)$ corresponds to the normal distribution. Clearly $\phi(s)\phi(-s)$ also has the form $\exp(P(t))$, where $P(s)$ is a polynomial that has only even terms, ie. $P(s) = \sum_{k=0}^{n} a_k s^{2k}$. Since $\phi(s)\phi(-s) = |\phi(s)|^2$ is a real number for all s, the coefficients $a_1, \ldots, a_n$ of polynomial $P(\cdot)$ are real. Moreover, the n-th coefficient satisfies $a_n = -\gamma^2 < 0$, as the inequality $|\phi(t)| \leq 1$ holds for arbitrarily large real t. Therefore, taking $z = N\exp(i\pi/(2n))$, we obtain

$$|\phi(z)| \geq \exp(N(\gamma^2 - \epsilon(N))) \tag{2.13}$$

for large enough real N, where $\epsilon(N) \to 0$ as $N \to \infty$. On the other hand, using the explicit representation by expected value, we get

$$|\phi(z)| = |E\exp(izX)| \leq E\exp(N\sin(\pi/(2n))X)$$

$$= \phi(N\sin(\pi/(2n))) = \exp(P(N\sin(\pi/(2n))))$$

$$\leq \exp(N\sin(\pi/(2n))(\gamma^2 + \epsilon(N))).$$

As $N \to \infty$ the last inequality contradicts (2.13), unless $\sin(\pi/(2n)) = 1$, ie. unless $n = 1$. This means that P is of degree 2 and, since $P(0) = 0$, we have $P(t) = -\gamma^2 t$ for all t. $\square$

2.6   Large deviations

Formula (2.3) shows that a multivariate normal distribution is uniquely determined by the vector $\mathbf{m}$ of expected values and the covariance matrix $\mathbf{C}$. However, to compute probabilities of the events of interest might be quite difficult. As Theorem 2.2.7 shows, even writing explicitly the density is cumbersome in higher dimensions as it requires inverting large matrices. Additional difficulties arise in degenerate cases.

Here we shall present the logarithmic term in the asymptotic expansion for $P(\mathbf{X} \in nA)$ as $n \to \infty$. This is the so called large deviation estimate; it becomes more accurate for less likely events. The main feature is that it has relatively simple form and applies to all events. Higher order expansions are more accurate but work for fairly regular sets $A \subset \mathbb{R}^d$ only.

Let us first define the conjugate "norm" to the RKHS seminorm $\| \cdot \|$ defined by (2.7).

$$\|\mathbf{y}\|_* = \sup_{\mathbf{x} \in \mathbf{R}^d, \, \|\mathbf{x}\|=1} \mathbf{x} \cdot \mathbf{y}.$$

The conjugate norm has all the properties of the norm except that it can attain value ∞. To see this, and also to have a more explicit expression, decompose $\mathbb{R}^d$ into the orthogonal sum of the null space of $\mathbf{A}$ and the range of $\mathbf{A}$: $\mathbb{R}^d = \mathcal{N}(\mathbf{A}) \oplus \mathcal{R}(\mathbf{A})$; here $\mathbf{A}$ is the symmetric matrix from (2.4). Since $\mathbf{A} : \mathbb{R}^d \to \mathcal{R}(\mathbf{A})$ is onto, there is a right-inverse $\mathbf{A}^{-1} : \mathcal{R}(\mathbf{A}) \to \mathcal{R}(\mathbf{A}) \subset \mathbb{R}^d$.

For $\mathbf{y} \in \mathcal{R}(\mathbf{A})$ we have

$$\sup_{\|\mathbf{Ax}\|=1} \mathbf{x} \cdot \mathbf{y} = \sup_{\|\mathbf{Ax}\|=1} \mathbf{x} \cdot \mathbf{AA}^{-1}\mathbf{y} = \sup_{\|\mathbf{Ax}\|=1} \mathbf{A}^T\mathbf{x} \cdot \mathbf{A}^{-1}\mathbf{y} \tag{2.14}$$

Since $\mathbf{A}$ is symmetric and $\mathbf{A}^{-1}\mathbf{y} \in \mathcal{R}(\mathbf{A})$, for $\mathbf{y} \in \mathcal{R}(\mathbf{A})$ we have by (2.14)

$$\|\mathbf{y}\|_* = \sup_{\mathbf{x} \in \mathcal{R}(\mathbf{A}), \, \|\mathbf{x}\|=1} \mathbf{x} \cdot \mathbf{A}^{-1}\mathbf{y} = \|\mathbf{A}^{-1}\mathbf{y}\|.$$

For $\mathbf{y} \notin \mathcal{R}(\mathbf{A})$ we write $\mathbf{y} = \mathbf{y}_\mathcal{N} + \mathbf{y}_\mathcal{R}$, where $0 \neq \mathbf{y}_\mathcal{N} \in \mathcal{N}(\mathbf{A})$. Then we have $\sup_{\|\mathbf{Ax}\|=1} \mathbf{x} \cdot \mathbf{y} \geq \sup_{\mathbf{x} \in \mathcal{N}(\mathbf{A})} \mathbf{x} \cdot \mathbf{y}_\mathcal{N} = \infty$. Since $\mathbf{C} = \mathbf{A} \times \mathbf{A}$, we get

$$\|\mathbf{y}\|_* = \begin{cases} \mathbf{y} \cdot \mathbf{C}^{-1}\mathbf{y} & \text{if } \mathbf{y} \in \mathcal{R}(\mathbf{C}); \\ \infty & \text{if } \mathbf{y} \notin \mathcal{R}(\mathbf{C}), \end{cases} \tag{2.15}$$

where $\mathbf{C}^{-1}$ is the right inverse of the covariance matrix $\mathbf{C}$.

In this notation, the multivariate normal density is

$$f(\mathbf{x}) = Ce^{-\frac{1}{2}\|\mathbf{x}-\mathbf{m}\|_*^2}, \tag{2.16}$$

where C is the normalizing constant and the integration has to be taken over the Lebesgue measure λ on the support $\text{supp}(\mathbf{X}) = \{\mathbf{x} : \|\mathbf{x}\|_* < \infty\}$.

To state the Large Deviation Principle, by A° we denote the interior of a Borel subset $A \subset \mathbb{R}^d$.

Theorem 2.6.1 *If* $\mathbf{X}$ *is Gaussian* $\mathbb{R}^d$*-valued with the mean* $\mathbf{m}$ *and the covariance matrix* $\mathbf{C}$, *then for all measurable* $A \subset \mathbb{R}^d$

$$\limsup_{n\to\infty} \frac{1}{n^2} \log P(\mathbf{X} \in nA) \leq - \inf_{\mathbf{x}\in A} \frac{1}{2} \|\mathbf{x} - \mathbf{m}\|_*^2 \qquad (2.17)$$

and

$$\liminf_{n\to\infty} \frac{1}{n^2} \log P(\mathbf{X} \in nA) \geq - \inf_{\mathbf{x}\in A^\circ} \frac{1}{2} \|\mathbf{x} - \mathbf{m}\|_*^2. \qquad (2.18)$$

The usual interpretation is that the dominant term in the asymptotic expansion for $P(\frac{1}{n}\mathbf{X} \in A)$ as $n \to \infty$ is given by

$$\exp\left(-\frac{n^2}{2} \inf_{\mathbf{x}\in A} \|\mathbf{x} - \mathbf{m}\|_*^2\right).$$

Proof. Clearly, passing to $\mathbf{X} - \mathbf{m}$ we can easily reduce the question to the centered random vector $\mathbf{X}$. Therefore we assume

$$\mathbf{m} = 0.$$

Inequality (2.17) follows immediately from

$$P(\mathbf{X} \in nA) = \int_{\mathrm{supp}(\mathbf{X})\cap A} Cn^{-k} e^{-\frac{n^2}{2}\|\mathbf{x}\|_*^2} \, d\mathbf{x}$$

$$\leq Cn^{-k}\lambda(\mathrm{supp}(\mathbf{X}) \cap A) \sup_{\mathbf{x}\in A} e^{-\frac{n^2}{2}\|\mathbf{x}\|_*^2},$$

where $C = C(k)$ is the normalizing constant and $k \leq d$ is the dimension of $\mathrm{supp}(\mathbf{X})$, cf. (2.16). Indeed,

$$\frac{1}{n^2} \log P(\mathbf{X} \in nA) \leq \frac{C}{n^2} - k\frac{\log n}{n^2} + \frac{\log \lambda(\mathrm{supp}(\mathbf{X}) \cap A)}{n^2} - \frac{1}{2} \inf_{\mathbf{x}\in A} \|\mathbf{x}\|_*^2.$$

To prove inequality (2.18) without loss of generality we restrict our attention to open sets A. Let $\mathbf{x}_0 \in A$. Then for all $\epsilon > 0$ small enough, the balls $B(\mathbf{x}_0, \epsilon) = \{\mathbf{x} : \|\mathbf{x} - \mathbf{x}_0\| < \epsilon\}$ are in A. Therefore

$$P(\mathbf{X} \in nA) \geq P(\mathbf{X} \in nD_\epsilon) = \int_{D_\epsilon} Cn^{-k} e^{-\frac{n^2}{2}\|\mathbf{x}\|_*^2} \, d\mathbf{x}, \qquad (2.19)$$

where $D_\epsilon = B(\mathbf{x}_0, \epsilon) \cap \mathrm{supp}(\mathbf{X})$. On the support $\mathrm{supp}(\mathbf{X})$ the function $\mathbf{x} \mapsto \|\mathbf{x}\|_*$ is finite and convex; thus it is continuous. For every $\eta > 0$ one can find ϵ such that $\|\mathbf{x}\|_*^2 \geq \|\mathbf{x}_0\|_*^2 - \eta$ for all $\mathbf{x} \in D_\epsilon$. Therefore (2.19) gives

$$P(\mathbf{X} \in nA) \geq Cn^{-k} e^{-(1-\eta)\frac{n^2}{2}\|\mathbf{x}\|_*^2},$$

which after passing to the logarithms ends the proof. $\square$

Large deviation bounds for Gaussian vectors valued in infinite dimensional spaces and for Gaussian stochastic processes have similar form and involve the conjugate RKHS norm; needless to say, the proof that uses the density cannot go through; for the general theory of large deviations the reader is referred to [32].

2.6.1 A numerical example

Consider a bivariate normal (X, Y) with the covariance matrix $\begin{bmatrix} 1 & 1 \\ 1 & 2 \end{bmatrix}$. The conjugate

RKHS norm is then $\left\| \begin{bmatrix} x \\ y \end{bmatrix} \right\|_* = 2x^2 - 2xy + y^2$ and the corresponding unit ball is the ellipse

$2x^2 - 2xy + y^2 = 1$. Figure 2.1 illustrates the fact that one can actually *see* the conjugated RKHS norm. Asymptotic shapes in more complicated systems are more mysterious, see [127].

Figure 2.1: A sample of $N = 1500$ points from bivariate normal distribution.

2.7 Problems

Problem 2.1 *If Z is the standard normal $N(0,1)$ random variable, show by direct integration that its characteristic function is $\phi(z) = \exp(-\frac{1}{2}z^2)$ for all complex $z \in \mathbf{C}$.*

Problem 2.2 *Suppose $(\mathbf{X}, \mathbf{Y}) \in \mathbb{R}^{d_1 + d_2}$ are jointly normal and have pairwise uncorrelated components, $\mathrm{corr}(X_i, Y_j) = 0$. Show that $\mathbf{X}, \mathbf{Y}$ are independent.*

Problem 2.3 *For standardized bivariate normal X, Y with correlation coefficient ρ, show that $P(X > 0, Y > 0) = \frac{1}{4} + \frac{1}{2\pi} \arcsin \rho$.*

Problem 2.4 *Prove Theorem 2.2.6.*

Problem 2.5 *Prove that "moments" $m_k = E\{X^k \exp(-X^2)\}$ are finite and determine the distribution of X uniquely.*

Problem 2.6 *Show that the exponential distribution is determined uniquely by its moments.*

Problem 2.7 *If $\phi(s)$ is an analytic characteristic function, show that $\log \phi(ix)$ is a well defined convex function of the real argument x.*

Problem 2.8 (deterministic analogue of Theorem 2.5.2) *Suppose ϕ_1, ϕ_2 are characteristic functions such that $\phi_1(t)\phi_2(t) = \exp(it)$ for each $t \in \mathbb{R}$. Show that $\phi_k(t) = \exp(ita_k)$, $k = 1, 2$, where $a_1, a_2 \in \mathbb{R}$.*

Problem 2.9 (exponential analogue of Theorem 2.5.2) *If X, Y are i. i. d. random variables such that $\min\{X, Y\}$ has an exponential distribution, then X is exponential.*

Chapter 3

Equidistributed linear forms

In Section 1.1 we present the classical characterization of the normal distribution by stability. Then we use this to define *Gaussian measures* on abstract spaces and we prove the zero-one law. In Section 3.3 we return to the characterizations of normal distributions. We consider a more difficult problem of characterizations by the equality of distributions of two general linear forms.

3.1 Two-stability

The main result of this section is the theorem due to G. Polya [122]. Polya's result was obtained before the axiomatization of probability theory. It was stated in terms of positive integrable functions and part of the conclusion was that the integrals of those functions are one, so that indeed the probabilistic interpretation is valid.

Theorem 3.1.1 *If X_1, X_2 are two i. i. d. random variables such that X_1 and $(X_1 + X_2)/\sqrt{2}$ have the same distribution, then X_1 is normal.*

It is easy to see that if X_1 and X_2 are i. i. d. random variables with the distribution corresponding to the characteristic function $\exp(-|t|^p)$, then the distributions of X_1 and $(X_1 + X_2)/\sqrt[p]{2}$ are equal. In particular, if X_1, X_2 are normal N(0,1), then so is $(X_1 + X_2)/\sqrt{2}$. Theorem 3.1.1 says that the above trivial implication can be inverted for $p = 2$. Corresponding results are also known for $p < 2$, but in general there is no uniqueness, see [133, 134, 135]. For $p \neq 2$ it is not obvious whether $\exp(-|t|^p)$ is indeed a characteristic function; in fact this is true only if $0 \leq p \leq 2$; the easier part of this statement was given as Problem 1.18. The distributions with this characteristic function are the so called (symmetric) stable distributions.

The following corollary shows that p-stable distributions with $p < 2$ cannot have finite second moments.

Corollary 3.1.2 *Suppose X_1, X_2 are i. i. d. random variables with finite second moments and such that for some scale factor κ and some location parameter α the distribution of $X_1 + X_2$ is the same as the distribution of $\kappa(X_1 + \alpha)$. Then X_1 is normal.*

Indeed, subtracting the expected value if necessary, we may assume $EX_1 = 0$ and hence $\alpha = 0$. Then $Var(X_1 + X_2) = Var(X_1) + Var(X_2)$ gives $\kappa = 2^{-1/2}$ (except if $X_1 = 0$; but this by definition is normal, so there is nothing to prove). By Theorem 3.1.1, X_1 (and also X_2) is normal.

Proof of Theorem 3.1.1. Clearly the assumption of Theorem 3.1.1 is not changed, if we pass to the symmetrizations $\widetilde{X}, \widetilde{Y}$ of X, Y. By Theorem 2.5.2 to prove the theorem, it remains to show that $\widetilde{X}$ is normal. Let $\phi(t)$ be the characteristic function of $\widetilde{X}, \widetilde{Y}$. Then

$$\phi(\sqrt{2}t) = \phi^2(t) \tag{3.1}$$

for all real t. Therefore recurrently we get

$$\phi(t2^{k/2}) = \phi(t)^{2^k} \tag{3.2}$$

for all real t. Take t_0 such that $\phi(t_0) \neq 0$; such t_0 can be found as ϕ is continuous and $\phi(0) = 1$. Let $\sigma^2 > 0$ such that $\phi(t_0) = \exp(-\sigma^2)$. Then (3.2) implies $\phi(t_0 2^{-k/2}) = \exp(-\sigma^2 2^{-k})$ for all $k = 0, 1, \ldots$. By Corollary 2.3.4 we have $\phi(t) = \exp(-\sigma^2 t^2)$ for all t, and the theorem is proved. $\square$

3.2   Measures on linear spaces

Let V be a linear space over the field $\mathbb{R}$ of real numbers (we shall also call V a (real) vector space). Suppose V is equipped with a σ-field $\mathcal{F}$ such that the algebraic operations of scalar multiplication $(\mathbf{x}, t) \mapsto t\mathbf{x}$ and of vector addition $\mathbf{x}, \mathbf{y} \mapsto \mathbf{x} + \mathbf{y}$ are measurable transformations $\mathsf{V} \times \mathbb{R} \to \mathsf{V}$ and $\mathsf{V} \times \mathsf{V} \to \mathsf{V}$ with respect to the corresponding σ-fields $\mathcal{F} \otimes \mathcal{B}_{\mathbb{R}}$, and $\mathcal{F} \otimes \mathcal{F}$ respectively. Let $(\Omega, \mathcal{M}, P)$ be a probability space. A measurable function $\mathbf{X} : \Omega \to \mathsf{V}$ is called a V-valued random variable.

Example 3.2.1 *Let $\mathsf{V} = \mathbb{R}^d$ be the vector space of all real d-tuples with the usual Borel σ-field $\mathcal{B}$. A V-valued random variable is called a d-dimensional random vector. Clearly $\mathbf{X} = (X_1, \ldots, X_d)$ and if one prefers, one can consider the family $X_1, \ldots, X_d$ rather than $\mathbf{X}$.*

Example 3.2.2 *Let $\mathsf{V} = C[0, 1]$ be the vector space of all continuous functions $[0, 1] \to \mathbb{R}$ with the topology defined by the norm $\|f\| := \sup_{0 \le t \le 1} |f(t)|$ and with the σ-field $\mathcal{F}$ generated by all open sets. Then a V-valued random variable $\mathbf{X}$ is called a stochastic process with continuous trajectories with time $T = [0, 1]$. The usual form is to write $X(t)$ for the random continuous function $\mathbf{X}$ evaluated at a point $t \in [0, 1]$.*

Warning. Although it is known that every abstract random vector can be interpreted as a random process with the appropriate choice of time set T, the natural choice of T (such as $T = 1, 2, \ldots, d$ in Example 3.2.1 and $T = [0, 1]$ in Example 3.2.2) might sometimes fail. For instance, let $\mathsf{V} = L_2[0, 1]$ be the vector space of all (classes of equivalence) of square integrable functions $[0, 1] \to \mathbb{R}$ with the usual L_2 norm $\|f\| = (\int f^2(t)\, dt)^{1/2}$. In general, a V-valued random variable $\mathbf{X}$ cannot be represented as a stochastic process with time $T = [0, 1]$, because evaluation at a point $t \in T$ is not a well defined mapping. Although $L_2[0, 1]$ is commonly thought as *the square integrable functions*, we are actually dealing with the classes of equivalence rather than with the genuine functions. For $V = L_2[0,1]$-valued Gaussian processes, one can show that X_t exists almost surely as the limit in probability of continuous linear functionals; abstract variants of this result can be found in [146] and in the references therein.

The following definition of an *abstract* Gaussian random variable is motivated by Theorem 3.1.1.

Definition 3.2.1 *A V-valued random variable $\mathbf{X}$ is $\mathcal{E}$-Gaussian ($\mathcal{E}$ stays for the equality of distributions) if the distribution of $\sqrt{2}\mathbf{X}$ is equal to the distribution of $\mathbf{X} + \mathbf{X}'$, where $\mathbf{X}'$ is an independent copy of $\mathbf{X}$.*

In Sections 5.2 and 5.4 we shall see that there are other equally natural candidates for the definitions of a Gaussian vector. To distinguish between them, we shall keep the longer name $\mathcal{E}$-Gaussian instead of just calling it *Gaussian*. Fortunately, at least in familiar situations, it does not matter which definition we use. This occurs whenever we have plenty of measurable linear functionals. By Theorem 3.1.1 if $\mathcal{L} : V \to \mathbb{R}$ is a measurable linear functional, then the $\mathbb{R}$-valued random variable $X = \mathcal{L}(\mathbf{X})$ is normal. When this specifies the probability measure on V uniquely, then all three definitions are equivalent. Let us see, how this works in two simple but important cases.

Example 3.2.1 (continued) *Suppose $\mathbf{X} = (X(1), X(2), \ldots, X(n))$ is an $\mathbb{R}^n$-valued $\mathcal{E}$-Gaussian random variable. Consider linear functionals $\mathcal{L} : \mathbb{R}^n \to \mathbb{R}$ given by $\mathcal{L}\mathbf{x} \mapsto \sum a_i x_i$, where $a_1, a_2, \ldots, a_n \in \mathbb{R}$. Then the one-dimensional random variable $a_1\mathbf{X}(1) + a_2\mathbf{X}(2) + \ldots + a_n\mathbf{X}(n)$ has the normal distribution. This means that $\mathbf{X}$ is a Gaussian vector in the usual sense (ie. it has multivariate normal distribution), as presented in Section 2.2.*

Example 3.2.2 (continued) *Suppose $\mathbf{X}$ is a $C[0,1]$-valued Gaussian random variable. Consider the set of all linear functionals $\mathcal{L} : C[0,1] \to \mathbb{R}$ that can be written in the form*

$$\mathcal{L} = a_1 \mathcal{E}_{t(1)} + a_2 \mathcal{E}_{t(2)} + \ldots + a_n \mathcal{E}_{t(n)},$$

where $a_1, \ldots, a_n$ are real numbers and $\mathcal{E}_t : C[0,1] \to \mathbb{R}$ denotes the evaluation at point t defined by $\mathcal{E}_t(f) = f(t)$. Then $\mathcal{L}(\mathbf{X}) = \sum a_i X(t_i)$ is normal. However, since the coefficients $a_1, \ldots, a_n$ are arbitrary, this means that for each choice of $t_1, t_2, \ldots, t_n \in [0,1]$ the n-dimensional random variable $X(t_1), X(t_2), \ldots, X(t_n)$ has a multivariate normal distribution, ie. $X(t)$ is a Gaussian stochastic process in the usual sense[1].

The question that we want to address now is motivated by the following (false) intuition. Suppose a measurable linear subspace $\mathbb{L} \subset V$ is given. Think for instance about $\mathbb{L} = C_1[0,1]$ – the space of all continuously differentiable functions, considered as a subspace of $C[0,1] = V$. In general, it seems plausible that some of the realizations of a V-valued random variable $\mathbf{X}$ may happen to fall in $\mathbb{L}$, while other realizations fail to be in $\mathbb{L}$. In other words, it seems plausible that with positive probability some of the trajectories of a stochastic process with continuous trajectories are smooth, while other

---

[1] In general, a family of T-indexed random variables $X(t)_{t \in T}$ is called a *Gaussian process* on T, if for every $n \geq 1$, $t_1, \ldots, t_n \in T$ the n-dimensional random vector $(X(t_1), \ldots, X(t_n))$ has multivariate normal distribution.

trajectories are not. Strangely, this cannot happen for Gaussian vectors (and, more generally, for α-stable vectors). The result is due to Dudley and Kanter and provides an example of the so called zero-one law. The most famous zero-one law is of course the one due to Kolmogorov, see eg. [9, Theorem 22.3]; see also the appendix to [82, page 69]. The proof given below follows [55]. Smoleński [138] gives an elementary proof, which applies also to other classes of measures. Krakowiak [89] proves the zero-one law when $\mathbb{L}$ is a measurable sub-group rather than a measurable linear subspace. Tortrat [143] considers (among other issues) zero-one laws for Gaussian distributions on groups. Theorem 5.2.1 and Theorem 5.4.1 in the next chapter give the same conclusion under different definitions of the Gaussian random vector.

Theorem 3.2.1 *If* $\mathbf{X}$ *is a* V*-valued* $\mathcal{E}$*-Gaussian random variable and* $\mathbb{L}$ *is a linear measurable subspace of* V, *then* $P(\mathbf{X} \in \mathbb{L})$ *is either 0, or 1.*

Proof. Let $\mathbf{X}_1, \mathbf{X}_2, \ldots$ be independent copies of $\mathbf{X}$. Also, let us choose them to be independent of $\mathbf{X}$. By 2-stability and the linearity of $\mathbb{L}$ we have

$$P(\mathbf{X}_1 + \mathbf{X}_2 \in \mathbb{L}) = P(\sqrt{2}\mathbf{X} \in \mathbb{L}) = P(\mathbf{X} \in \mathbb{L}). \tag{3.3}$$

By induction, this gives

$$P(\mathbf{X}_1 + \mathbf{X}_2 + \ldots + \mathbf{X}_{2^n} \in \mathbb{L}) = P(\mathbf{X} \in \mathbb{L}) \tag{3.4}$$

for all $n = 0, 1, \ldots$.

Let $\mathbf{Z} = \mathbf{X}_1 + \mathbf{X}_2$. Clearly, $\mathbf{Z}$ is independent of $\mathbf{X}$ and 2-stability implies that $\mathbf{X}_1 + \mathbf{X}_2 + \ldots + X_{2^n+1}$ has the same distribution as $\mathbf{Z} + 2^{n/2}\mathbf{X}$. Therefore (3.4) gives

$$P(\mathbf{Z} + 2^{n/2}\mathbf{X} \in \mathbb{L}) = P(\mathbf{X} \in \mathbb{L}). \tag{3.5}$$

Consider now events $A_n = \{\mathbf{Z} \notin \mathbb{L}\} \cap \{\mathbf{Z} + 2^{n/2}\mathbf{X} \in \mathbb{L}\}$. Since event $\{\mathbf{Z} \in \mathbb{L}\} \cap \{\mathbf{Z} + 2^{n/2}X \in \mathbb{L}\}$ is the same as $\{\mathbf{Z} \in \mathbb{L}\} \cap \{\mathbf{X} \in \mathbb{L}\}$, therefore by (3.5)

$$P(A_n) = P(\mathbf{Z} + 2^{n/2}\mathbf{X} \in \mathbb{L}) - P(\mathbf{Z} \in \mathbb{L})P(\mathbf{X} \in \mathbb{L})$$

$$= P(\mathbf{X} \in \mathbb{L}) - P(\mathbf{Z} \in \mathbb{L})P(\mathbf{X} \in \mathbb{L}).$$

By (3.3) this says that $P(A_n) = P(\mathbf{X} \in \mathbb{L})P(\mathbf{X} \notin \mathbb{L})$ does not depend on n.

Now let us observe that if $m \neq n$, then the events A_m and A_n are disjoint. We shall prove this by contradiction. Suppose both vectors $\mathbf{Z} + 2^{n/2}\mathbf{X} \in \mathbb{L}$ and $\mathbf{Z} + 2^{m/2}\mathbf{X} \in \mathbb{L}$. Then their difference $(2^{n/2} - 2^{m/2})\mathbf{X}$ is in $\mathbb{L}$, too. For $m \neq n$ this implies $\mathbf{X} \in \mathbb{L}$ and therefore $\mathbf{Z} \in \mathbb{L}$. The latter contradicts the definition of A_n, proving that A_m and A_n are indeed disjoint.

The preceding two observations show that $\{A_n\}$ is an infinite sequence of disjoint events with the same probability fixed $P(A_n) = P(A_1)$. This can happen only if $P(A_n) = 0$, ie. when $P(\mathbf{X} \in \mathbb{L})P(\mathbf{X} \notin \mathbb{L}) = 0$, which ends the proof. $\square$

To make Theorem 3.2.1 more concrete, consider the following application.

Example 3.2.3 *This example presents a simple-minded model of transmission of information. Suppose that we have a choice of one of the two signals $f(t)$, or $g(t)$ be transmitted by a noisy channel within unit time interval $0 \leq t \leq 1$. To simplify the situation even further, we assume $g(t) = 0$, ie. g represents "no message send". The noise (which is always present) is a random and continuous function; we shall assume that it is represented by a $C[0,1]$-valued Gaussian random variable $W = \{W(t)\}_{0 \leq t \leq 1}$. We also assume it is an "additive" noise.*

Under these circumstances the signal received is given by a curve; it is either $\{f(t) + W(t)\}_{0 \leq t \leq 1}$, or $\{W(t)\}_{0 \leq t \leq 1}$, depending on which of the two signals, f or g, was sent. The objective is to use the received signal to decide, which of the two possible messages: $f(\cdot)$ or 0 (ie. message, or no message) was sent.

Notice that, at least from the mathematical point of view, the task is trivial if $f(\cdot)$ is known to be discontinuous; then we only need to observe the trajectory of the received signal and check for discontinuities. There are of course numerous practical obstacles to collecting continuous data, which we are not going to discuss here.

If $f(\cdot)$ is continuous, then the above procedure does not apply. Problem requires more detailed analysis in this case. One may adopt the usual approach of testing the null hypothesis that no signal was sent. This amounts to choosing a suitable critical region $\mathbb{L} \subset C[0,1]$. As usual in statistics, the decision is to be made according to whether the observed trajectory falls into $\mathbb{L}$ (in which case we decide $f(\cdot)$ was sent) or not (in which case we decide that 0 was sent and that what we have received was just the noise). Clearly, to get a sensible test we need $P(f(\cdot) + W(\cdot) \in \mathbb{L}) > 0$ and $P(W(\cdot) \in \mathbb{L}) < 1$.

Theorem 3.2.1 implies that perfect discrimination is achieved if we manage to pick the critical region in the form of a (measurable) linear subspace. Indeed, then by Theorem 3.2.1 $P(W(\cdot) \in \mathbb{L}) < 1$ implies $P(W(\cdot) \in \mathbb{L}) = 0$ and $P(f(\cdot) + W(\cdot) \in \mathbb{L}) > 0$ implies $P(f(\cdot) + W(\cdot) \in \mathbb{L}) = 1$.

Unfortunately, it is not true that a linear space can always be chosen for the critical region. For instance, if $W(\cdot)$ is the Wiener process (see Section 8.1), it is known that such subspace <u>cannot</u> be found if (and only if!) $f(\cdot)$ is differentiable for almost all t and $\int(\frac{df}{dt})^2 \, dt < \infty$. The proof of this theorem is beyond the scope of this book (cf. Cameron-Martin formula in [41]). The result, however, is surprising (at least for those readers, who know that trajectories of the Wiener process are non-differentiable): it implies that, at least in principle, each non-differentiable (everywhere) signal $f(\cdot)$ can be recognized without errors despite having non-differentiable Wiener noise.

(Affine subspaces for centered noise $EW_t = 0$ do not work, see Problem 3.4)

For a recent work, see [44].

3.3   Linear forms

It is easily seen that if $a_1, \ldots, a_n$ and $b_1, \ldots, b_n$ are real numbers such that the sets $A = \{|a_1|, \ldots, |a_n|\}$ and $B = \{|b_1|, \ldots, |b_n|\}$ are equal, then for any symmetric i. i. d. random variables $X_1, \ldots, X_n$ the sums $\sum_{k=1}^{n} a_k X_k$ and $\sum_{k=1}^{n} b_k X_k$ have the same distribution. On the other hand, when $n = 2$, $A = \{1, 1\}$ and $B = \{0, \sqrt{2}\}$ Theorem 3.1.1 says that the equality of distributions of linear forms $\sum_{k=1}^{n} a_k X_k$ and $\sum_{k=1}^{n} b_k X_k$ implies normality. In this section we shall consider two more characterizations of the normal distribution by the equality of distributions of linear combinations $\sum_{k=1}^{n} a_k X_k$ and $\sum_{k=1}^{n} b_k X_k$. The results are considerably less elementary than Theorem 3.1.1.

We shall begin with the following generalization of Corollary 3.1.2 which we learned from J. Wesołowski.

Theorem 3.3.1 *Let $X_1, \ldots, X_n, n \geq 2$, be i. i. d. square-integrable random variables and let $A = \{a_1, \ldots, a_n\}$ be the set of real numbers such that $A \neq \{1, 0, \ldots, 0\}$. If X_1 and $\sum_{k=1}^{n} a_k X_k$ have equal distributions, then X_1 is normal.*

The next lemma is a variant of the result due to C. R. Rao, see [73, Lemma 1.5.10].

Lemma 3.3.2 *Suppose $q(\cdot)$ is continuous in a neighborhood of 0, $q(0) = 0$, and in a neighborhood of 0 it satisfies the equation*

$$q(t) = \sum_{k=1}^{n} a_k^2 q(a_k t), \tag{3.6}$$

where $a_1, \ldots, a_n$ are given numbers such that $|a_k| \leq \delta < 1$ and $\sum_{k=1}^{n} a_k^2 = 1$.
Then $q(t) = const$ in some neighborhood of $t = 0$.

Proof. Suppose (3.6) holds for all $|t| < \epsilon$. Then $|a_j t| < \epsilon$ and from (3.6) we get $q(a_j t) = \sum_{k=1}^{n} a_k^2 q(a_j a_k t)$ for every $1 \leq j \leq n$. Hence $q(t) = \sum_{j=1}^{n} \sum_{k=1}^{n} a_j^2 a_k^2 q(a_j a_k t)$ and we get recurrently

$$q(t) = \sum_{j_1=1}^{n} \cdots \sum_{j_r=1}^{n} a_{j_1}^2 \cdots a_{j_r}^2 q(a_{j_1} \cdots a_{j_r} t)$$

for all $r \geq 1$. This implies

$$|q(t) - q(0)| \leq \left(\sum_{k=1}^{n} a_k^2 \right)^r \sup_{|a| \leq \delta^r} |q(at) - q(0)| = \sup_{|x| \leq \delta^r} |q(x) - q(0)| \to 0$$

as $r \to \infty$ for all $|t| < \epsilon$. $\square$

Proof of Theorem 3.3.1. Without loss of generality we may assume $Var(X_1) \neq 0$. Let ϕ be the characteristic function of X and let $Q(t) = \log \phi(t)$. Clearly, $Q(t)$ is well defined for all t close enough to 0. Equality of distributions gives

$$Q(t) = Q(a_1 t) + Q(a_2 t) + \ldots + Q(a_n t).$$

The integrability assumption implies that Q has two derivatives, and for all t close enough to 0 the derivative $q(\cdot) = Q''(\cdot)$ satisfies equation (3.6).

Since X_1 and $\sum_{k=1}^{n} a_k X_k$ have equal variances, $\sum_{k=1}^{n} a_k^2 = 1$. Condition $|a_i| \neq 0, 1$ implies $|a_i| < 1$ for all $1 \leq i \leq n$. Lemma 3.3.2 shows that $q(\cdot)$ is constant in a neighborhood of $t = 0$ and ends the proof. $\square$

Comparing Theorems 3.1.1 and 3.3.1 the pattern seems to be that the less information about coefficients, the more information about the moments is needed. The next result ([106]) fits into this pattern, too; [73, Section 2.3 and 2.4] present the general theory of *active exponents* which permits to recognize (by examining the coefficients of linear forms), when the equality of distributions of linear forms implies normality; see also [74]. Variants of characterizations by equality of distributions are known for group-valued random variables, see [50]; [49] is also pertinent.

Theorem 3.3.3 *Suppose $A = \{|a_1|, \ldots, |a_n|\}$ and $B = \{|b_1|, \ldots, |b_n|\}$ are different sets of real numbers and $X_1, \ldots, X_n$ are i. i. d. random variables with finite moments of all orders. If the linear forms $\sum_{k=1}^{n} a_k X_k$ and $\sum_{k=1}^{n} b_k X_k$ are identically distributed, then X_1 is normal.*

We shall need the following elementary lemma.

Lemma 3.3.4 *Suppose $A = \{|a_1|, \ldots, |a_n|\}$ and $B = \{|b_1|, \ldots, |b_n|\}$ are different sets of real numbers. Then*

$$(\sum_{k=1}^{n} a_k^{2r}) \neq (\sum_{k=1}^{n} b_k^{2r}) \tag{3.7}$$

for all $r \geq 1$ large enough.

Proof. Without loss of generality we may assume that coefficients are arranged in increasing order $|a_1| \leq \ldots \leq |a_n|$ and $|b_1| \leq \ldots \leq |b_n|$. Let M be the largest number $m \leq n$ such that $|a_m| \neq |b_m|$. (Clearly, at least one such m exists, because sets A, B consist of different numbers.) Then $|a_k| = |b_k|$ for $k > M$ and $\sum_{k=1}^{n} a_k^{2r} \neq \sum_{k=1}^{n} b_k^{2r}$ for all r large enough. Indeed, by the definition of M we have $\sum_{k>M} b_k^{2r} = \sum_{k>M} a_k^{2r}$ but the remaining portions of the sum are not equal, $\sum_{k \leq M} b_k^{2r} \neq \sum_{k \leq M} a_k^{2r}$ for r large enough; the latter holds true because by our choice of M the limits $\lim_{r \to \infty} (\sum_{k \leq M} a_k^{2r})^{1/(2r)} = \max_{k \leq M} |a_k| = |a_M|$ and $\lim_{r \to \infty} (\sum_{k \leq M} b_k^{2r})^{1/(2r)} = \max_{k \leq M} |b_k| = |b_M|$ are not equal. $\square$

We also need the following lemma[2] due to Marcinkiewicz [106].

Lemma 3.3.5 *Let ϕ be an infinitely differentiable characteristic function and let $Q(t) = \log \phi(t)$. If there is $r \geq 1$ such that $Q^{(k)}(0) = 0$ for all $k \geq r$, then ϕ is the characteristic function of a normal distribution.*

Proof. Indeed, $\Phi(z) = \exp(\sum_{k=0}^{r} \frac{z^k}{k!} Q^{(k)}(0))$ is an analytic function and all derivatives at 0 of the functions $\log \Phi(\cdot)$ and $\log \phi(\cdot)$ are equal. Differentiating the (trivial) equality $\phi Q' = \phi'$, we get $\phi^{(n+1)} = \sum_{k=0}^{n} \binom{n}{k} \phi^{(n-k)} Q^{(k+1)}$, which shows that all derivatives at 0 of $\Phi(\cdot)$ and of $\phi(\cdot)$ are equal. This means that $\phi(\cdot)$ is analytic in some neighborhood of 0 and $\phi(t) = \Phi(t) = \exp P(t)$ for all small enough t, where P is a polynomial of the degree (at most) r. Hence by Theorem 2.5.3, ϕ is normal. $\square$

Proof of Theorem 3.3.3. Without loss of generality, we may assume that X_1 is symmetric. Indeed, if random variables $X_1, \ldots, X_n$ satisfy the assumptions of the theorem, then so do their symmetrizations $\widetilde{X}_1, \ldots, \widetilde{X}_n$, see Section 1.6. If we could prove the theorem for symmetric random variables, then $\widetilde{X}_1$ would be be normal. By Theorem 2.5.2, this would

---

[2]For a recent application of this lemma to the Central Limit Problem, see [68].

imply that X_1 is normal. Hence it suffices to prove the theorem under the additional symmetry assumption. Let ϕ be the characteristic function of X's and let $Q(t) = \log \phi(t); Q$ is well defined for all t close enough to 0. The assumption implies that Q has derivatives of all orders and also that $Q(a_1 t) + Q(a_2 t) + \ldots + Q(a_n t) = Q(b_1 t) + Q(b_2 t) + \ldots + Q(b_n t)$. Differentiating the last equality $2r$ times at $t = 0$ we obtain

$$\sum_{k=1}^{n} a_k^{2r} Q^{(2r)}(0) = \sum_{k=1}^{n} b_k^{2r} Q^{(2r)}(0), r = 0, 1, \ldots \tag{3.8}$$

Notice that by (3.7), equality (3.8) implies $Q^{(2r)}(0) = 0$ for all r large enough. Thus by (3.8) (and by the symmetry assumption to handle the derivatives of odd order), $Q^{(k)}(0) = 0$ for all $k \geq 1$ large enough. Lemma 3.3.5 ends the proof. $\square$

3.4   Exponential analogy

Characterizations of the normal distribution frequently lead to analogous characterizations of the exponential distribution. The idea behind this correspondence is that adding random variables is replaced by taking their minimum. This is explained by the well known fact that the minimum of independent exponential random variables is exponentially distributed; the observation is due to Linnik [100], see [73, p. 87]. Monographs [57, 4], present such results as well as the characterizations of the exponential distribution by its intrinsic properties, such as *lack of memory*. In this book some of the exponential analogues serve as exercises.

The following result, written in the form analogous to Theorem 0.0.1, illustrates how the exponential analogy works. The i. i. d. assumption can easily be weakened to independence of X and Y (the details of this modification are left to the reader as an exercise).

Theorem 3.4.1 *Suppose X, Y non-negative random variables such that*
(i) for all $a, b > 0$ such that $a + b = 1$, the random variable $\min\{X/a, Y/b\}$ has the same distribution as X;
(ii) X and Y are independent and identically distributed.
Then X and Y are exponential.

Proof. The following simple observation stays behind the proof.

If X, Y are independent non-negative random variables, then the tail distribution function, defined for any $Z \geq 0$ by $N_Z(x) = P(Z \geq x)$, satisfies

$$N_{\min\{X,Y\}}(x) = N_X(x) N_Y(x). \tag{3.9}$$

Using (3.9) and the assumption we obtain $N(at)N(bt) = N(t)$ for all $a, b, t > 0$ such that $a + b = 1$. Writing $t = x + y, a = x/(x + y), b = y/(x + y)$ for arbitrary $x, y > 0$ we get

$$N(x + y) = N(x) N(y) \tag{3.10}$$

Therefore to prove the theorem, we need only to solve functional equation (3.10) for the unknown function $N(\cdot)$ such that $0 \leq N(\cdot) \leq 1$; $N(\cdot)$ is also right-continuous non-increasing and $N(x) \to 0$ as $x \to \infty$.

Formula (3.10) shows recurrently that for all integer n and all $x \geq 0$ we have

$$N(nx) = N(x)^n. \tag{3.11}$$

Since $N(0) = 1$ and $N(\cdot)$ is right continuous, it follows from (3.11) that $r = N(1) > 0$. Therefore (3.11) implies $N(n) = r^n$ and $N(1/n) = r^{1/n}$ (to see this, plug in (3.11) values $x = 1$ and $x = 1/n$ respectively). Hence $N(n/m) = N(1/m)^n = r^{n/m}$ (by putting $x = 1/m$ in (3.11)), ie. for each rational $q > 0$ we have

$$N(q) = r^q. \tag{3.12}$$

Since $N(x)$ is right-continuous, $N(x) = \lim_{q \searrow x} N(q) = r^x$ for each $x \geq 0$. It remains to notice that $r < 1$, which follows from the fact that $N(x) \to 0$ as $x \to \infty$. Therefore $r = \exp(-\lambda)$ for some $\lambda > 0$, and $N(x) = \exp(-\lambda x), x \geq 0$. $\square$

3.5 Exponential distributions on lattices

The abstract notation of this section follows [43, page 43]. Let $\mathbb{L}$ be a vector space with norm $\|\cdot\|$. Suppose that $\mathbb{L}$ is also a lattice with the operations *minimum* $\wedge$ and *maximum* $\vee$ which are consistent with the vector operations and with the norm. The related order is then defined by $\mathbf{x} \preceq \mathbf{y}$ iff $\mathbf{x} \vee \mathbf{y} = \mathbf{y}$ (or, alternatively: iff $\mathbf{x} \wedge \mathbf{y} = \mathbf{x}$). By consistency with vector operations we mean that[3]

$$(\mathbf{x} + \mathbf{y}) \wedge (\mathbf{z} + \mathbf{y}) = \mathbf{y} + (\mathbf{x} \wedge \mathbf{z}) \text{ for all } \mathbf{x}, \mathbf{y}, \mathbf{z} \in \mathbb{L}$$

$$(\alpha \mathbf{x}) \wedge (\alpha \mathbf{y}) = \alpha(\mathbf{x} \wedge \mathbf{y}) \text{ for all } \mathbf{x}, \mathbf{y} \in \mathbb{L}, \alpha \geq 0$$

and

$$-(\mathbf{x} \wedge \mathbf{y}) = (-\mathbf{x}) \vee (-\mathbf{y}).$$

Consistency with the norm means

$$\|\mathbf{x}\| \leq \|\mathbf{y}\| \text{ for all } 0 \preceq \mathbf{x} \preceq \mathbf{y}$$

Moreover, we assume that there is a σ-field $\mathcal{F}$ such that all the operations considered are measurable.

Vector space $\mathbb{R}^d$ with

$$\mathbf{x} \wedge \mathbf{y} = (\min\{x_j; y_j\})_{1 \leq j \leq d} \tag{3.13}$$

with the norm: $\|\mathbf{x}\| = \max_j |x_j|$ satisfies the above requirements. Other examples are provided by the function spaces with the usual norms; for instance, a familiar example is the space $C[0,1]$ of all continuous functions with the standard supremum norm and the pointwise minimum of functions as the lattice operation, is a lattice.

The following abstract definition complements [57, Chapter 5].

---

[3]See eg. [43, page 43] or [3].

Definition 3.5.1 *A random variable* $\mathbf{X} : \Omega \to \mathbb{L}$ *has exponential distribution if the following two conditions are satisfied: (i)* $\mathbf{X} \succeq 0$;

(ii) if $\mathbf{X}'$ *is an independent copy of* $\mathbf{X}$ *then for any* $0 < a < 1$ *random variables* $\mathbf{X}/a \wedge \mathbf{X}'/(1-a)$ *and* $\mathbf{X}$ *have the same distribution.*

Example 3.5.1 *Let* $\mathbb{L} = \mathbb{R}^d$ *with* $\wedge$ *defined coordinatewise by (3.13) as in the above discussion. Then any* $\mathbb{R}^d$*-valued exponential random variable has the multivariate exponential distribution in the sense of Pickands, see [57, Theorem 5.3.7]. This distribution is also known as Marshall-Olkin distribution.*

Using the definition above, it is easy to notice that if $(X_1, \ldots, X_d)$ has the exponential distribution, then $\min\{X_1, \ldots, X_d\}$ has the exponential distribution on the real line. The next result is attributed to Pickands see [57, Section 5.3].

Proposition 3.5.1 *Let* $\mathbf{X} = (X_1, \ldots, X_d)$ *be an* $\mathbb{R}^d$*-valued exponential random variable. Then the real random variable* $\min\{X_1/a_1, \ldots, X_d/a_d\}$ *is exponential for all* $a_1, \ldots, a_d > 0$.

Proof. Let $Z = \min\{X_1/a_1, \ldots, X_d/a_d\}$. Let Z' be an independent copy of Z. By Theorem 3.4.1 it remains to show that

$$\min\{Z/a; Z'/b\} \cong Z \tag{3.14}$$

for all $a, b > 0$ such that $a + b = 1$. It is easily seen that

$$\min\{Z/a; Z'/b\} = \min\{Y_1/a_1, \ldots, Y_d/a_d\},$$

where $Y_i = \min\{X_i/a; X_i'/b\}$ and $\mathbf{X}'$ is an independent copy of $\mathbf{X}$. However by the definition, $\mathbf{X}$ has the same distribution as $(Y_1, \ldots, Y_d)$, so (3.14) holds. $\square$

Remark: By taking a limit as $a_j \to 0$ for all $j \neq i$, from Proposition 3.5.1 we obtain in particular that each component X_i is exponential.

Example 3.5.2 *Let* $\mathbb{L} = C[0,1]$ *with* $\{f \wedge g\}(x) := \min\{f(x), g(x)\}$. *Then exponential random variable* $\mathbf{X}$ *defines the stochastic process* $X(t)$ *with continuous trajectories and such that* $\{X(t_1), X(t_2), \ldots, X(t_n)\}$ *has the n-dimensional Marshall-Olkin distribution for each integer n and for all* $t_1, \ldots, t_n$ *in* $[0,1]$.

The following result shows that the supremum $\sup_t |X(t)|$ of the *exponential process* from Example 3.5.2 has the moment generating function in a neighborhood of 0. Corresponding result for Gaussian processes will be proved in Sections 5.2 and 5.4. Another result on *infinite dimensional* exponential distributions will be given in Theorem 4.3.4.

Proposition 3.5.2 *If* $\mathbb{L}$ *is a lattice with the measurable norm* $\|\cdot\|$ *consistent with algebraic operation* $\wedge$, *then for each exponential* $\mathbb{L}$*-valued random variable* $\mathbf{X}$ *there is* $\lambda > 0$ *such that* $E\exp(\lambda\|\mathbf{X}\|) < \infty$.

Proof. The result follows easily from the trivial inequality

$$P(\|\mathbf{X}\| \geq 2x) = P(\|\mathbf{X} \wedge \mathbf{X}'\| \geq x) \leq (P(\|\mathbf{X}\| \geq x))^2$$

and Corollary 1.3.7. $\square$

3.6 Problems

Problem 3.1 (deterministic analogue of Theorem 3.1.1)) *Show that if $X, Y \geq 0$ are i. i. d. and $2X$ has the same distribution as $X + Y$, then X, Y are non-random* [4].

Problem 3.2 *Suppose random variables X_1, X_2 satisfy the assumptions of Theorem 3.1.1 and have finite second moments. Use the Central Limit Theorem to prove that X_1 is normal.*

Problem 3.3 *Let V be a metric space with a measurable metric d. We shall say that a V-valued sequence of random variables S_n converges to Y in distribution, if there exist a sequence $\hat{S}_n$ convergent to Y in probability (ie. $P(d(\hat{S}_n, Y) > \epsilon) \to 0$ as $n \to \infty$) and such that $S_n \cong \hat{S}_n$ (in distribution) for each n. Let $\mathbf{X}_n$ be a sequence of V-valued independent random variables and put $S_n = \mathbf{X}_1 + \ldots + \mathbf{X}_n$. Show that if S_n converges in distribution (in the above sense), then the limit is an $\mathcal{E}$-Gaussian random variable* [5].

Problem 3.4 *For a separable Banach-space valued Gaussian vector X define the mean $\mathbf{m} = E\mathbf{X}$ as the unique vector that satisfies $\lambda(\mathbf{m}) = E\lambda(\mathbf{X})$ for all continuous linear functionals $\lambda \in \mathsf{V}^\star$. It is also known that random vectors with equal characteristic functions $\phi(\lambda) = E \exp i\lambda(\mathbf{X})$ have the same probability distribution.*

Suppose $\mathbf{X}$ is a Gaussian vector with the non-zero mean $\mathbf{m}$. Show that for a measurable linear subspace $\mathbb{L} \subset \mathsf{V}$, if $\mathbf{m} \notin \mathbb{L}$ then $P(\mathbf{X} \in \mathbb{L}) = 0$.

Problem 3.5 (deterministic analogue of Theorem 3.3.2)) *Show that if i. i. d. random variables X, Y have moments of all orders and $X + 2Y \cong 3X$, then X, Y are non-random.*

Problem 3.6 *Show that if X, Y are independent and $X + Y \cong X$, then $Y = 0$ a. s.*

---

[4] Cauchy distribution shows that assumption $X \geq 0$ is essential.
[5] For motivation behind such a definition of weak convergence, see Skorohod [137].

Chapter 4

Rotation invariant distributions

4.1 Spherically symmetric vectors

Definition 4.1.1 *A random vector* $\mathbf{X} = (X_1, X_2, \ldots, X_n)$ *is spherically symmetric if the distribution of every linear form*

$$a_1 X_1 + a_2 X_2 + \ldots + a_n X_n \cong X_1 \tag{4.1}$$

is the same for all $a_1, a_2, \ldots, a_n$, *provided* $a_1^2 + a_2^2 + \ldots + a_n^2 = 1$.

A slightly more general class of the so called *elliptically contoured distributions* has been studied from the point of view of applications to statistics in [47]. Elliptically contoured distributions are images of spherically symmetric random variables under a linear transformation of $\mathbb{R}^n$. Additional information can also be found in [48, Chapter 4], which is devoted to the characterization problems and overlaps slightly with the contents of this section.

Let $\phi(\mathbf{t})$ be the characteristic function of $\mathbf{X}$. Then

$$\phi(\mathbf{t}) = \phi\left(\|\mathbf{t}\| \begin{bmatrix} 1 \\ 0 \\ \vdots \\ 0 \end{bmatrix} \right), \tag{4.2}$$

ie. the characteristic function at $\mathbf{t}$ can be written as a function of $\|\mathbf{t}\|$ only. Conversely, if $\phi(t)$ is a characteristic function of a real random variable, then $\phi(\|\mathbf{t}\|)$ corresponds to an $\mathbb{R}^n$-valued random vector.

From the definition we also get the following.

Proposition 4.1.1 *If* $\mathbf{X} = (X_1, \ldots, X_n)$ *is spherically symmetric, then each of its marginals* $\mathbf{Y} = (X_1, \ldots, X_k)$, *where* $k \leq n$, *is spherically symmetric.*

This fact is very simple; just consider linear forms (4.1) with $a_{k+1} = \ldots = a_n = 0$.

Example 4.1.1 *Suppose* $\vec{\gamma} = (\gamma_1, \gamma_2, \ldots, \gamma_n)$ *is the sequence of independent identically distributed normal* $N(0, 1)$ *random variables. Then* $\vec{\gamma}$ *is spherically symmetric. Moreover, for any* $m \geq 1, \vec{\gamma}$ *can be extended to a longer spherically invariant sequence*

$(\gamma_1, \gamma_2, \ldots, \gamma_{n+m})$. *In Theorem 4.3.1 we will see that up to a random scaling factor, this is essentially the only example of a spherically symmetric sequence with arbitrarily long spherically symmetric extensions*[1].

In general a multivariate normal distribution is not spherically symmetric. But if $\mathbf{X}$ is centered non-degenerated Gaussian r. v., then $\mathbf{A}^{-1}\mathbf{X}$ is spherically symmetric, see Theorem 2.2.4. Spherical symmetry together with Theorem 4.1.2 is sometimes useful in computations as illustrated in Problem 4.2.

Example 4.1.2 *Suppose* $\mathbf{X} = (X_1, \ldots, X_n)$ *has the uniform distribution on the sphere* $\|\mathbf{x}\| = r$. *Obviously,* $\mathbf{X}$ *is spherically symmetric. For* $k < n$, *vector* $\mathbf{Y} = (X_1, \ldots, X_k)$ *has the density*

$$f(\mathbf{y}) = C(r^2 - \|\mathbf{y}\|^2)^{(n-k)/2-1}, \tag{4.3}$$

where C is the normalizing constant (see for instance, [48, formula (1.2.6)]). In particular, $\mathbf{Y}$ *is spherically symmetric and absolutely continuous in* $\mathbb{R}^k$.

The density of real valued random variable $Z = \|\mathbf{Y}\|$ *at point z has an additional factor coming from the area of the sphere of radius z in* $\mathbb{R}^k$, *ie.*

$$f_Z(z) = C z^{k-1}(r^2 - z^2)^{(n-k)/2-1}. \tag{4.4}$$

Here $C = C(r, k, n)$ is again the normalizing constant. By rescaling, it is easy to see that $C = r^{n-2}C_1(k, n)$, *where*

$$C_1(k, n) = (\int_{-1}^{1} z^{k-1}(1 - z^2)^{(n-k)/2-1}\, dz)^{-1}$$

$$= \frac{2\Gamma(n/2)}{\Gamma(k/2)\Gamma((n-k)/2)} = \frac{2}{B(k/2, (n-k)/2)}.$$

Therefore

$$f_Z(z) = C_1 r^{n-2} z^{k-1}(r^2 - z^2)^{(n-k)/2-1}. \tag{4.5}$$

Finally, let us point out that the conditional distribution of $\|(X_{k+1}, \ldots, X_n)\|$ *given* $\mathbf{Y}$ *is concentrated at one point* $(r^2 - \|\mathbf{Y}\|^2)^{1/2}$.

From expression (4.3) it is easy to see that for fixed k, if $n \to \infty$ and the radius is $r = \sqrt{n}$, then the density of the corresponding $\mathbf{Y}$ converges to the density of the i. i. d. normal sequence $(\gamma_1, \gamma_2, \ldots, \gamma_k)$. (This well known fact is usually attributed to H. Poincaré).

Calculus formulas of Example 4.1.2 are important for the general spherically symmetric case because of the following representation.

Theorem 4.1.2 *Suppose* $\mathbf{X} = (X_1, \ldots, X_n)$ *is spherically symmetric. Then* $\mathbf{X} = R\mathbf{U}$, *where random variable* $\mathbf{U}$ *is uniformly distributed on the unit sphere in* $\mathbb{R}^n$, $R \geq 0$ *is real valued with distribution* $R \cong \|\mathbf{X}\|$, *and random variables variables* $R, \mathbf{U}$ *are stochastically independent.*

---

[1]It would be interesting to find the conditions that allow to recognize when a given spherically symmetric random vector can be embedded as a marginal of a higher dimensional spherically symmetric one.

Proof. The first step of the proof is to show that the distribution of $\mathbf{X}$ is invariant under all rotations $\mathbf{U} : \mathbb{R}^n \to \mathbb{R}^n$. Indeed, since by definition $\phi(t) = E\exp(it \cdot \mathbf{X}) = E\exp(i\|t\|X_1)$, the characteristic function $\phi(t)$ of $\mathbf{X}$ is a function of $\|t\|$ only. Therefore the characteristic function ψ of $\mathbf{UX}$ satisfies

$$\psi(t) = E\exp(it \cdot \mathbf{UX}) = E\exp(i\mathbf{U}^T t \cdot \mathbf{X}) = E\exp(i\|t\|X_1) = \phi(t).$$

The group $O(n)$ of rotations of $\mathbb{R}^n$ (ie. the group of orthogonal $n \times n$ matrices) is a compact group; by μ we denote the normalized Haar measure (cf. [59, Section 58]). Let $\mathbf{G}$ be an $O(n)$-valued random variable with the distribution μ and independent of $\mathbf{X}$ ($\mathbf{G}$ can be actually written down explicitly; for example if $n = 2$, $\mathbf{G} = \begin{bmatrix} \cos\theta & \sin\theta \\ -\sin\theta & \cos\theta \end{bmatrix}$, where θ is uniformly distributed on $[0, 2\pi]$.) Clearly $\mathbf{X} \cong \mathbf{GX} \cong \|\mathbf{X}\|\mathbf{GX}/\|\mathbf{X}\|$ conditionally on the event $\|\mathbf{X}\| \neq 0$. To take care of the possibility that $\mathbf{X} = 0$, let $\mathbf{\Theta}$ be uniformly distributed on the unit sphere and put

$$\mathbf{U} = \begin{cases} \mathbf{\Theta} & \text{if } \mathbf{X} = 0 \\ \mathbf{GX}/\|\mathbf{X}\| & \text{if } \mathbf{X} \neq 0 \end{cases}.$$

It is easy to see that $\mathbf{U}$ is uniformly distributed on the unit sphere in $\mathbb{R}^n$ and that $\mathbf{U}, \mathbf{X}$ are independent. This ends the proof, since $\mathbf{X} \cong \mathbf{GX} = \|\mathbf{X}\|\mathbf{U}$. $\square$

The next result explains the connection between spherical symmetry and linearity of regression. Actually, condition (4.6) under additional assumptions characterizes elliptically contoured distributions, see [61, 118].

Proposition 4.1.3 *If $\mathbf{X}$ is a spherically symmetric random vector with finite first moments, then*

$$E\{X_1|a_1X_1 + \ldots + a_nX_n\} = \rho \sum_{k=1}^{n} a_k X_k \tag{4.6}$$

for all real numbers $a_1, \ldots, a_n$, where $\rho = \frac{a_1}{a_1^2 + \ldots + a_n^2}$.

Sketch of the proof.
The simplest approach here is to use the converse to Theorem 1.5.3; if $\phi(\|t\|^2)$ denotes the characteristic function of $\mathbf{X}$ (see (4.2)), then the characteristic function of $X_1, a_1X_1 + \ldots + a_nX_n$ evaluated at point (t, s) is $\psi(t, s) = \phi((s + a_1t)^2 + (a_2t)^2 + \ldots + (a_nt)^2)$. Hence

$$(a_1^2 + \ldots + a_n^2)\frac{\partial}{\partial s}\psi(t, s)\bigg|_{s=0} = a_1\frac{\partial}{\partial t}\psi(t, 0).$$

Another possible proof is to use Theorem 4.1.2 to reduce (4.7) to the uniform case. This can be done as follows. Using the well known properties of conditional expectations, we have

$$E\{X_1|a_1X_1 + \ldots + a_nX_n\} = E\{RU_1|R(a_1U_1 + \ldots + a_nU_n)\}$$

$$= E\{E\{RU_1|R, a_1U_1 + \ldots + a_nU_n\}|R(a_1U_1 + \ldots + a_nU_n)\}.$$

Clearly,

$$E\{RU_1|R, a_1U_1 + \ldots + a_nU_n\} = RE\{U_1|R, a_1U_1 + \ldots + a_nU_n\}$$

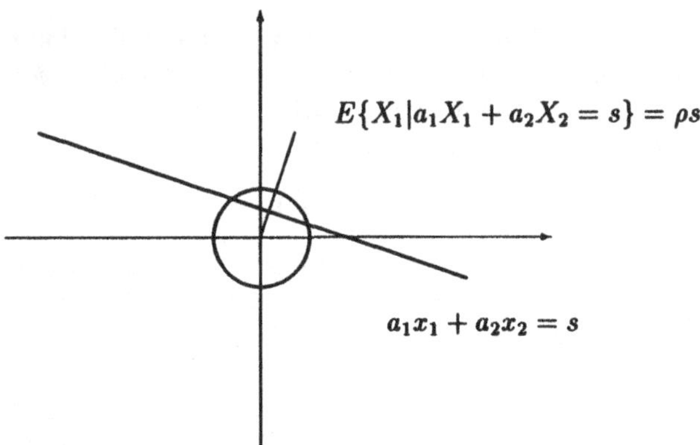

Figure 4.1: Linear regression for the uniform distribution on a circle.

and

$$E\{U_1|R, a_1U_1 + \ldots + a_nU_n\} = E\{U_1|a_1U_1 + \ldots + a_nU_n\},$$

see Theorem 1.4.1 (ii) and (iii). Therefore it suffices to establish (4.7) for the uniform distribution on the unit sphere. The last fact is quite obvious from symmetry considerations; for the 2-dimensional situation this can be illustrated on a picture. Namely, the hyper-plane $a_1x_1 + \ldots + a_nx_n = const$ intersects the unit sphere along a translation of a suitable $(n-1)$-dimensional sphere S; integrating x_1 over S we get the same fraction (which depends on $a_1, \ldots, a_n$) of $const$. $\square$

The following theorem shows that spherical symmetry allows us to eliminate the assumption of independence in Theorem 0.0.1, see also Theorem 7.2.1 below. The result for rational α is due to S. Cambanis, S. Huang & G. Simons [25]; for related exponential results see [57, Theorem 2.3.3].

Theorem 4.1.4 *Let* $\mathbf{X} = (X_1, \ldots, X_n)$ *be a spherically symmetric random vector such that* $E\|\mathbf{X}\|^\alpha < \infty$ *for some real* $\alpha > 0$. *If*

$$E\{\|(X_1, \ldots, X_m)\|^\alpha|(X_{m+1}, \ldots, X_n)\} = const$$

for some $1 \le m < n$, *then* $\mathbf{X}$ *is Gaussian.*

Our method of proof of Theorem 4.1.4 will also provide easy access to the following interesting result due to Szabłowski [140, Theorem 2], see also [141].

Theorem 4.1.5 *Let* $\mathbf{X} = (X_1, \ldots, X_n)$ *be a spherically symmetric random vector such that* $E\|\mathbf{X}\|^2 < \infty$ *and* $P(\mathbf{X} = 0) = 0$. *Suppose* $c(x)$ *is a real function with the property that there is* $0 \le U \le \infty$ *such that* $1/c(x)$ *is integrable on each finite sub-interval of the interval* $[0, U]$ *and that* $c(x) = 0$ *for all* $x > U$.

If for some $1 \le m < n$

$$E\{\|(X_1, \ldots, X_m)\|^2|(X_{m+1}, \ldots, X_n)\} = c(\|(X_{m+1}, \ldots, X_n)\|),$$

then the distribution of $\mathbf{X}$ *is determined uniquely by* $c(x)$.

To prove both theorems we shall need the following.

Lemma 4.1.6 *Let* $\mathbf{X} = (X_1, \ldots, X_n)$ *be a spherically symmetric random vector such that* $P(\mathbf{X} = 0) = 0$ *and let* H *denote the distribution of* $\|\mathbf{X}\|$. *Then we have the following.*

(a) For $m < n$ *r. v.* $\|(X_{m+1}, \ldots, X_n)\|$ *has the density function* $g(x)$ *given by*

$$g(x) = Cx^{n-m-1} \int_x^\infty r^{-n+2}(r^2 - x^2)^{m/2-1} H(dr), \qquad (4.7)$$

where $C = 2\Gamma(\tfrac{1}{2}n)(\Gamma(\tfrac{1}{2}m)\Gamma(\tfrac{1}{2}(n-m)))^{-1}$ *is a normalizing constant of no further importance below.*

(b) The distribution of $\mathbf{X}$ *is determined uniquely by the distribution of its single component* X_1.

(c) The conditional distribution of $\|(X_1, \ldots, X_m)\|$ *given* $(X_{m+1}, \ldots, X_n)$ *depends only on the* $\mathbb{R}^{m-n}$*-norm* $\|(X_{m+1}, \ldots, X_n)\|$ *and*

$$E\{\|(X_1, \ldots, X_m)\|^\alpha | (X_{m+1}, \ldots, X_n)\} = h(\|(X_{m+1}, \ldots, X_n)\|),$$

where

$$h(x) = \frac{\int_x^\infty r^{-n+2}(r^2 - x^2)^{(m+\alpha)/2-1} H(dr)}{\int_x^\infty r^{-n+2}(r^2 - x^2)^{m/2-1} H(dr)} \qquad (4.8)$$

Sketch of the proof.
Formulas (4.7) and (4.8) follow from Theorem 4.1.2 by conditioning on R, see Example 4.1.2. Fact (b) seems to be intuitively obvious; it says that from the distribution of the product $U_1 R$ of independent random variables (where U_1 is the 1-dimensional marginal of the uniform distribution on the unit sphere in $\mathbb{R}^n$) we can recover the distribution of R. Indeed, this follows from Theorem 1.8.1 and (4.7) applied to $m = n - 1$: multiplying $g(x) = C \int_x^\infty r^{-n+2}(r^2 - x^2)^{(n-1)/2-1} H(dr)$ by x^{u-1} and integrating, we get the formula which shows that from $g(x)$ we can determine the integrals $\int_0^\infty r^{t-1} H(dr)$, cf. (4.10) below.$\square$

Lemma 4.1.7 *Suppose* $c_\alpha(\cdot)$ *is a function such that*

$$E\{\|(X_1, \ldots, X_m)\|^\alpha | (X_{m+1}, \ldots, X_n)\} = c_\alpha(\|(X_{m+1}, \ldots, X_n)\|^2).$$

Then the function $f(x) = x^{(m+1-n)/2}g(x^{1/2})$, *where* $g(.)$ *is defined by (4.7), satisfies*

$$c_\alpha(x)f(x) = \frac{1}{B(\alpha/2, m/2)} \int_x^\infty (y - x)^{\alpha/2-1} f(y)\, dy. \qquad (4.9)$$

Proof. As previously, let $H(dr)$ be the distribution of $\|\mathbf{X}\|$. The following formula for the beta integral is well known, cf. [110].

$$(r^2 - x^2)^{(m+\alpha)/2-1} = \frac{2}{B(\alpha/2, m/2)} \int_x^1 (t^2 - x^2)^{\alpha/2-1}(r^2 - t^2)^{m/2-1}\, dt. \qquad (4.10)$$

Substituting (4.10) into (4.8) and changing the order of integration we get

$$c_\alpha(x^2)g(x)$$

$$= C x^{n-m-1} \frac{2}{B(\alpha/2, m/2)} \int_x^\infty (t^2 - x^2)^{\alpha/2-1} \int_t^\infty r^{-n+2}(r^2 - t^2)^{m/2-1} H(dr)\, dt.$$

Using (4.7) we have therefore

$$c_\alpha(x^2) g(x) = x^{n-m-1} \frac{2}{B(\alpha/2, m/2)} \int_x^\infty (t^2 - r^2)^{\alpha/2-1} t^{m+1-n} g(t)\, dt.$$

Substituting $f(\cdot)$ and changing the variable of integration from t to t^2 ends the proof of (4.9). $\square$

Proof of Theorem 4.1.5. By Lemma 4.1.6 we need only to show that for $\alpha = 2$ equation (4.9) has the unique solution. Since $f(\cdot) \geq 0$, it follows from (4.9) that $f(y) = 0$ for all $y \geq U$. Therefore it suffices to show that $f(x)$ is determined uniquely for $x < U$. Since the right hand side of (4.9) is differentiable, therefore from (4.9) we get $2\frac{d}{dx}(c(x)f(x)) = -mf(x)$. Thus $\beta(x) := c(x)f(x)$ satisfies equation

$$2\beta'(x) = -m\beta(x)/c(x)$$

at each point $0 \leq x < U$. Hence $\beta(x) = C\exp(-\frac{1}{2}m \int_0^x 1/c(t)\, dt)$. This shows that

$$f(x) = \frac{C}{c(x)} \exp(-\frac{1}{2}m \int_0^x \frac{1}{c(t)}\, dt)$$

is determined uniquely (here $C > 0$ is a normalizing constant). $\square$

Lemma 4.1.8 *If $\pi(s)$ is a periodic and analytic function of complex argument s with the real period, and for real t the function $t \mapsto \log(\pi(t)\Gamma(t + C))$ is real valued and convex, then $\pi(s) = \text{const}$.*

Proof. For all positive x we have

$$\frac{d^2}{dx^2} \log \pi(x) + \frac{d^2}{dx^2} \log \Gamma(x) \geq 0. \tag{4.11}$$

However it is known that $\frac{d^2}{dx^2} \log \Gamma(x) = \sum_{n \geq 0}(n+x)^{-2} \to 0$ as $x \to \infty$, see [110]. Therefore (4.11) and the periodicity of $\pi(.)$ imply that $\frac{d^2}{dx^2} \log \pi(x) \geq 0$. This means that the first derivative $\frac{d}{dx} \log \pi(.)$ is a continuous, real valued, periodic and non-decreasing function of the real argument. Hence $\frac{d}{dx} \log \pi(x) = B \in \mathbb{R}$ for all real x. Therefore $\log \pi(s) = A + Bs$ and, since $\pi(.)$ is periodic with real period, this implies $B = 0$. This ends the proof. $\square$

Proof of Theorem 4.1.4. There is nothing to prove, if $\mathbf{X} = 0$. If $P(\mathbf{X} = 0) < 1$ then $P(\mathbf{X} = 0) = 0$. Indeed, suppose, on the contrary, that $P(\mathbf{X} = 0) > 0$. By Theorem 4.1.2 this means that $p = P(R = 0) > 0$ and that $E\{\|(X_1, \ldots, X_m)\|^\alpha | (X_{m+1}, \ldots, X_n)\} = 0$ with positive probability $p > 0$. Therefore $E\{\|(X_1, \ldots, X_m)\|^\alpha | (X_{m+1}, \ldots, X_n)\} = 0$ with probability 1. Hence $R = 0$ and $\mathbf{X} = 0$ a. s., a contradiction.

Throughout the rest of this proof we assume without loss of generality that $P(\mathbf{X} = 0) = 0$. By Lemmas 4.1.6 and 4.1.7, it remains to show that the integral equation

$$f(x) = K \int_x^\infty (y - x)^{\beta-1} f(y)\, dy \tag{4.12}$$

has the unique solution in the class of functions satisfying conditions $f(.) \geq 0$ and $\int_0^\infty x^{(n-m)/2-1} f(x)\, dx = 2$.

Let $\mathcal{M}(s) = x^{s-1} f(x) dx$ be the Mellin transform of $f(.)$, see Section 1.8. It can be checked that $\mathcal{M}(s)$ is well defined and analytic for s in the half-plane $\Re s > \frac{1}{2}(n - m)$, see Theorem 1.8.2. This holds true because the moments of all orders are finite, a claim which can be recovered with the help of a variant of Theorem 6.2.2, see Problem 6.6; for a stronger conclusion see also [22, Theorem 2.2]. The Mellin transform applied to both sides of (4.12) gives

$$\mathcal{M}(s) = K \frac{\Gamma(\beta)\Gamma(s)}{\Gamma(\beta + s)} \mathcal{M}(\beta + s).$$

Thus the Mellin transform $\mathcal{M}_1(.)$ of the function $f(Cx)$, where $C = (K\Gamma(\beta))^{-1/\beta}$, satisfies

$$\mathcal{M}_1(s) = \mathcal{M}_1(\beta + s) \frac{\Gamma(s)}{\Gamma(\beta + s)}.$$

This shows that $\mathcal{M}_1(s) = \pi(s)\Gamma(s)$, where $\pi(.)$ is analytic and periodic with real period β. Indeed, since $\Gamma(s) \neq 0$ for $\Re s > 0$, function $\pi(s) = \mathcal{M}_1(s)/\Gamma(s)$ is well defined and analytic in the half-plane $\Re s > 0$. Now notice that $\pi(.)$, being periodic, has analytic extension to the whole complex plane.

Since $f(.) \geq 0$, $\log \mathcal{M}_1(x)$ is a well defined *convex* function of the real argument x. This follows from the Cauchy-Schwarz inequality, which says that $\mathcal{M}_1((t + s)/2) \leq (\mathcal{M}_1(t)\mathcal{M}_1(s))^{1/2}$. Hence by Lemma 4.1.8, $\pi(s) = const.\square$

Remark: Solutions of equation (4.12) have been found in [62]. Integral equations of similar, but more general form occurred in potential theory, see Deny [33], see also Bochner [11] for an early work; for another proof and recent literature, see [126].

4.2   Rotation invariant absolute moments

The following beautiful theorem is due to M. S. Braverman [14][2].

Theorem 4.2.1 *Let X, Y, Z be independent identically distributed random variables with finite moments of fixed order $p \in \mathbb{R}_+ \setminus 2\mathbb{N}$. Suppose that there is constant C such that for all real a, b, c*

$$E|aX + bY + cZ|^p = C(a^2 + b^2 + c^2)^{p/2}. \tag{4.13}$$

Then X, Y, Z are normal.

---

[2]In the same paper Braverman also proves a similar characterization of α-stable distributions.

Condition (4.13) says that the absolute moments of a fixed order p of any axis, no matter how rotated, are the same; this fits well into the framework of Theorem 0.0.1.

Theorem 4.2.1 is a strictly 3-dimensional phenomenon, at least if no additional conditions on random variables are imposed. It does not hold for pairs of i. i. d. random variables, see Problem 4.3 below[3]. Theorem 4.2.1 cannot be extended to other values of exponent p; if p is an even integer, then (4.13) is not strong enough to imply the normal distribution (the easiest case to see this is of course $p = 2$).

Following Braverman's argument, we obtain Theorem 4.2.1 as a corollary to Theorem 3.1.1. To this end, we shall use the following result of independent interest.

Theorem 4.2.2 *If $p \in \mathbb{R}_+ \setminus 2\mathbb{N}$ and X, Y, Z are independent symmetric p-integrable random variables such that $P(Z = 0) < 1$ and*

$$E|X + tZ|^p = E|Y + tZ|^p \text{ for all real } t, \tag{4.14}$$

then $X \cong Y$ in distribution.

Theorem 4.2.2 resembles Problem 1.17, and it seems to be related to potential theory, see [123, page 65] and [80, Section 6]. Similar results have functional analytic importance, see Rudin [129]; also Hall [58] and Hardin [60] might be worth seeing in this context. Koldobskii [80, 81] gives Banach space versions of the results and relevant references.

Theorem 4.2.1 follows immediately from Theorem 4.2.2 by the following argument.
Proof of Theorem 4.2.1 . Clearly there is nothing to prove, if $C = 0$, see also Problem 4.5. Suppose therefore $C \neq 0$. It follows from the assumption that $E|X + Y + tZ|^p = E|\sqrt{2}X + tZ|^p$ for all real t. Note also that $E|Z|^p = C \neq 0$. Therefore Theorem 4.2.2 applied to $X + Y$, X' and Z, where X' is an independent copy of $\sqrt{2}X$, implies that $X + Y$ and $\sqrt{2}X$ have the same distribution. Since X, Y are i. i. d., by Theorem 3.1.1 X, Y, Z are normal. $\square$

A related result

The next result can be thought as a version of Theorem 4.2.1 corresponding to $p = 0$. For the proof see [85, 92, 96].

Theorem 4.2.3 *If $\mathbf{X} = (X_1, \ldots, X_n)$ is at least 3-dimensional random vector such that its components $X_1, \ldots, X_n$ are independent, $P(\mathbf{X} = 0) = 0$ and $\mathbf{X}/\|\mathbf{X}\|$ has the uniform distribution on the unit sphere in $\mathbb{R}^n$, then $\mathbf{X}$ is Gaussian.*

4.2.1   Proof of Theorem 4.2.2 for $p = 1$

We shall first present a slightly simplified proof for $p = 1$ which is based on elementary identity $\max\{x, y\} = (x + y + |x - y|)$. This proof leads directly to the exponential analogue of Theorem 4.2.1; the exponential version is given as Problem 4.4 below.

We shall begin with the lemma which gives an analytic version of condition (4.14).

---

[3]For more counter-examples, see also [15]; cf. also Theorems 4.2.8 and 4.2.9 below.

Lemma 4.2.4 *Let X_1, X_2, Y_1, Y_2 be symmetric independent random variables such that $E|Y_i| < \infty$ and $E|X_i| < \infty, i = 1, 2$. Denote $N_i(t) = P(|X_i| \geq t), M_i(t) = P(|Y_i| \geq t), t \geq 0, i = 1, 2$. Then each of the conditions*

$$E|a_1 X_1 + a_2 X_2| = E|a_1 Y_1 + a_2 Y_2| \text{ for all } a_1, a_2 \in \mathbb{R}; \tag{4.15}$$

$$\int_0^\infty N_1(\tau) N_2(x\tau)\, d\tau = \int_0^\infty M_1(\tau) M_2(x\tau)\, d\tau \text{ for all } x > 0; \tag{4.16}$$

$$\int_0^\infty N_1(xt) N_2(yt)\, dt = \int_0^\infty M_1(xt) M_2(yt)\, dt \tag{4.17}$$

$$\text{for all } x, y \geq 0, |x| + |y| \neq 0;$$

implies the other two.

Proof. For all real numbers x, y we have $|x - y| = 2\max\{x, y\} - (x + y)$. Therefore, taking into account the symmetry of the distributions for all real a, b we have

$$E|aX_1 - bX_2| = 2E\max\{aX_1, bX_2\}. \tag{4.18}$$

For an integrable random variable Z we have $EZ = \int_0^\infty P(Z \geq t)\, dt - \int_0^\infty P(-Z \geq t)\, dt$, see (1.3). This identity applied to $Z = \max\{aX_1, bX_2\}$, where $a, b \geq 0$ are fixed, gives

$$E\max\{aX_1, bX_2\} = \int_0^\infty P(Z \geq t)\, dt - \int_0^\infty P(Z \leq -t)\, dt$$

$$= \int_0^\infty P(aX_1 \geq t)\, dt + \int_0^\infty P(bX_2 \geq t)\, dt$$

$$- \int_0^\infty P(aX_1 \geq t)P(bX_2 \geq t)\, dt - \int_0^\infty P(aX_1 \leq -t)P(bX_2 \leq -t)\, dt.$$

Therefore, from (4.18) after taking the symmetry of distributions into account, we obtain

$$E|aX_1 - bX_2| = 2aEX_1^+ + 2bEX_2^+ - 4\int_0^\infty P(aX_1 \geq t)P(bX_2 \geq t)\, dt,$$

where $X_i^+ = \max\{X_i, 0\}, i = 1, 2$. This gives

$$E|aX_1 - bX_2| = 2aEX_1^+ + 2bEX_2^+ - 4\int_0^\infty N_1(t/a)N_2(t/b)\, dt. \tag{4.19}$$

Similarly

$$E|aY_1 - bY_2| = 2aEY_1^+ + 2bEY_2^+ - 4\int_0^\infty M_1(t/a)M_2(t/b)\, dt. \tag{4.20}$$

Once formulas (4.19) and (4.20) are established, we are ready to prove the equivalence of conditions (4.15)-(4.17).

(4.15)$\Rightarrow$(4.16): If condition (4.15) is satisfied, then $E|X_i| = E|Y_i|, i = 1, 2$ and thus by symmetry $EX_i^+ = EY_i^+, i = 1, 2$. Therefore (4.19) and (4.20) applied to $a = 1, b = 1/x$ imply (4.16) for any fixed $x > 0$.

(4.16) $\Rightarrow$(4.17): Changing the variable in (4.16) we obtain (4.17) for all $x > 0, y > 0$. Since $E|Y_i| < \infty$ and $E|X_i| < \infty$ we can pass in (4.17) to the limit as $x \to 0$, while y is fixed, or as $y \to 0$, while x is fixed, and hence (4.17) is proved in its full generality.

(4.17)$\Rightarrow$(4.15): If condition (4.17) is satisfied, then taking $x = 0, y = 1$ or $x = 1, y = 0$ we obtain $E|X_i| = E|Y_i|, i = 1, 2$ and thus by symmetry $EX_i^+ = EY_i^+, i = 1, 2$. Therefore identities (4.19) and (4.20) applied to $a = 1/x, b = 1/y$ imply (4.15) for any $a_1 > 0, a_2 < 0$. Since $E|Y_i| < \infty$ and $E|X_i| < \infty$, we can pass in (4.15) to the limit as $a_1 \to 0$, or as $a_2 \to 0$. This proves that equality (4.15) for all $a_1 \geq 0, a_2 \leq 0$. However, since X_i, Y_i, $i = 1, 2$, are symmetric, this proves (4.15) in its full generality. $\square$

The next result translates (4.15) into the property of the Mellin transform. A similar analytical identity is used in the proof of Theorem 4.2.3.

Lemma 4.2.5 *Let X_1, X_2, Y_1, Y_2 be symmetric independent random variables such that $E|Y_j| < \infty$ and $E|X_j| < \infty, j = 1, 2$. Let $0 < u < 1$ be fixed. Then condition (4.15) is equivalent to*

$$E|X_1|^{u+it}E|X_2|^{1-u-it} = E|Y_1|^{u+it}E|Y_2|^{1-u-it} \text{ for all } t \in \mathbb{R}. \tag{4.21}$$

Proof. By Lemma 2.4.3, it suffice to show that conditions (4.21) and (4.16) are equivalent.

Proof of (4.16)$\Rightarrow$(4.21): Multiplying both sides of (4.16) by x^{-u-it}, where $t \in \mathbb{R}$ is fixed, integrating with respect to x in the limits from 0 to ∞ and changing the order of integration (which is allowed, since the integrals are absolutely convergent), then substituting $x = y/\tau$, we get

$$\int_0^\infty \tau^{it+u-1} N_1(\tau)\, dt \int_0^\infty y^{-u-it} N_2(y)\, dy$$

$$= \int_0^\infty \tau^{it+u-1} M_1(\tau)\, d\tau \int_0^\infty y^{-u-it} M_2(y)\, dy.$$

This clearly implies (4.21), since, eg.

$$\int_0^\infty \tau^{it+u-1} N_j(\tau)\, d\tau = E|X_j|^{u+it}/(u+it), \ j = 1, 2$$

(this is just tail integration formula (1.2)).

Proof of (4.21)$\Rightarrow$(4.16): Notice that

$$\phi_j(t) := \frac{uE|X_j|^{u+it}}{(u+it)E|X_j|^u}, \ j = 1, 2$$

is the characteristic function of a random variable with the probability density function $f_{j,u}(x) := C_j \exp(xu)N_j(\exp(x))$, $x \in \mathbb{R}$, $j = 1, 2$, where $C_j = C_j(u)$ is the normalizing constant. Indeed,

$$\int_{-\infty}^\infty e^{ixt} \exp(xu)N_j(\exp(x))\, dx = \int_0^\infty y^{it}y^{u-1}N_j(y)\, dy = E|X_j|^{u+it}/(u+it)$$

and the normalizer $C_j(u) = u/E|X_j|^u$ is then chosen to have $\phi_j(0) = 1$, $j = 1, 2$. Similarly

$$\psi_j(t) := \frac{uE|Y_j|^{u+it}}{(u+it)E|Y_j|^u}$$

is the characteristic function of a random variable with the probability density function $g_{j,u}(x) := K_j \exp(xu)M_j(\exp(x)), x \in \mathbb{R}$, where $K_j = u/E|Y_j|^u, j = 1, 2$. Therefore (4.21) implies that the following two convolutions are equal $f_{1,u} * \bar{f}_{2,1-u} = g_{1,u} * \bar{g}_{2,1-u}$, where $\bar{f}_2(x) = f_2(-x), \bar{g}_2(x) = g_2(-x)$. Since (4.21) implies $C_1(u)C_2(1-u) = K_1(u)K_2(1-u)$, a simple calculation shows that the equality of convolutions implies

$$\int_{-\infty}^{\infty} e^x N_1(e^x) N_2(e^y e^x) \, dx = \int_{-\infty}^{\infty} e^x M_1(e^x) M_2(e^y e^x) \, dx$$

for all real y. The last equality differs from (4.16) by the change of variable only. $\square$

Now we are ready to prove Theorem 4.2.2. The conclusion of Lemma 4.2.5 suggests using the Mellin transform $E|X|^{u+it}, t \in \mathbb{R}$. Recall from Section 1.8 that if for some fixed $u > 0$ we have $E|X|^u < \infty$, then the function $E|X|^{u+it}$, $t \in \mathbb{R}$, determines the distribution of $|X|$ uniquely. This and Lemma 4.2.5 are used in the proof of Theorem 4.2.2.

Proof of Theorem 4.2.2. Lemma 4.2.5 implies that for each $0 < u < 1, -\infty < t < \infty$

$$E|X|^{u+it} E|Z|^{1-u-it} = E|Y|^{u+it} E|Z|^{1-u-it}. \tag{4.22}$$

Since $E|Z|^s$ is an analytic function in the strip $0 < \Re s < 1$, see Theorem 1.8.2, and $E|Z| = C \neq 0$ by (4.13), therefore the equation $E|Z|^{u+it} = 0$ has at most a countable number of solutions (u, t) in the strip $0 < u < 1$ and $-\infty < t < \infty$. Indeed, the equation has at most a finite number of solutions in each compact set — otherwise we would have $Z = 0$ almost surely by the uniqueness of analytic extension. Therefore one can find $0 < u < 1$ such that $E|Z|^{u+it} \neq 0$ for all $t \in \mathbb{R}$. For this value of u from (4.22) we obtain

$$E|X|^{1-u-it} = E|Y|^{1-u-it} \tag{4.23}$$

for all real t, which by Theorem 1.8.1 proves that random variables X and Y have the same distribution. $\square$

4.2.2 Proof of Theorem 4.2.2 in the general case

The following lemma shows that under assumption (4.14) all even moments of order less than p match.

Lemma 4.2.6 *Let $k = [p/2]$. Then (4.14) implies*

$$E|X|^{2j} = E|Y|^{2j} \tag{4.24}$$

for $j = 0, 1, \ldots, k$.

Proof. For $j \leq k$ the derivatives $\frac{\partial^j}{\partial t^j}|tX + Z|^p$ are integrable. Therefore (4.24) follows by the consecutive differentiation (under the integral signs) of the equation $E|tX + Z|^p = E|tY + Z|^p$ at $t = 0$. $\square$

The following is a general version of (4.21).

Lemma 4.2.7 *Let $0 < u < p$ be fixed. Then condition (4.14) and*

$$E|X|^{u+it}E|Z|^{p-u-it} = E|Y|^{u+it}E|Z|^{p-u-it} \text{ for all } t \in \mathbb{R}. \tag{4.25}$$

are equivalent.

Proof. We prove only the implication (4.14)$\Rightarrow$(4.25); we will not use the other one.

Let $k = [p/2]$. The following elementary formula follows by the change of variable[4]

$$|a|^p = C_p \int_0^\infty \left(\cos ax - \sum_{j=0}^{k} (-1)^j a^{2j} x^{2j} \right) \frac{dx}{x^{p+1}} \tag{4.26}$$

for all a.

Since our variables are symmetric, applying (4.26) to $a = X + \alpha Z$ and $a = Y + \alpha Z$ from (4.14) and Lemma 4.2.6 we get

$$\int_0^\infty \frac{(\phi_X(x) - \phi_Y(x))\phi_Z(\alpha x)}{x^{p+1}} \, dx = 0 \tag{4.27}$$

and the integral converges absolutely. Multiplying (4.27) by $\alpha^{-p+u+it-1}$, integrating with respect to α in the limits from 0 to ∞ and switching the order of integrals we get

$$\int_0^\infty \frac{\phi_X(x) - \phi_Y(x)}{x^{p+1}} \int_0^\infty \alpha^{-p+u+it-1} \phi_Z(\alpha x) \, d\alpha \, dx = 0. \tag{4.28}$$

Notice that

$$\int_0^\infty \alpha^{-p+u+it-1} \phi_Z(\alpha x) \, d\alpha = x^{p-u-it} \int_0^\infty \beta^{-p+u+it-1} \phi_Z(\beta) \, d\beta$$

$$= x^{p-u-it}\Gamma(-p + u + it)E|Z|^{p-u-it}.$$

Therefore (4.28) implies

$$\Gamma(-p + u + it)\Gamma(-u - it)\left(E|X|^{u+it} - E|Y|^{u+it} \right) E|Z|^{p-u-it} = 0.$$

This shows that identity (4.25) holds for all values of t, except perhaps a for a countable discrete set arising from the zeros of the Gamma function. Since $E|Y|^z$ is analytic in the strip $-1 < \Re z < p$, this implies (4.25) for all t. $\square$

Proof of Theorem 4.2.2 (general case). The proof of the general case follows the previous argument for $p = 1$ with (4.25) replacing (4.21). $\square$

---

[4]Notice that our choice of k ensures integrability.

4.2.3 Pairs of random variables

Although in general Theorem 4.2.1 doesn't hold for a pair of i. i. d. variables, it is possible to obtain a variant for pairs under additional assumptions. Braverman [16] obtained the following result.

Theorem 4.2.8 *Suppose X, Y are i. i. d. and there are positive $p_1 \neq p_2$ such that $p_1, p_2 \notin 2\mathbb{N}$ and $E|aX + bY|^{p_j} = C_j(a^2 + b^2)^{p_j}$ for all $a, b \in \mathbb{R}$, $j = 1, 2$. Then X is normal.*

Proof of Theorem 4.2.8. Suppose $0 < p_1 < p_2$. Denote by Z the standard normal N(0,1) random variable and let

$$f_p(s) = \frac{E|X|^{p/2+s}}{E|Z|^{p/2+s}}.$$

Clearly f_p is analytic in the strip $-1 < p/2 + \Re s < p_2$.

For $-p_1/2 < \Re s < p_2/2$ by Lemma 4.2.7 we have

$$f_{p_1}(s)f_{p_1}(-s) = C_1 \tag{4.29}$$

and

$$f_{p_2}(s)f_{p_2}(-s) = C_2 \tag{4.30}$$

Put $r = \frac{1}{2}(p_2 - p_1)$. Then $f_{p_2}(s) = f_{p_1}(s + r)$ in the strip $-p_1/2 < \Re s < p_1/2$. Therefore (4.30) implies

$$f(r + s)f(r - s) = C_2,$$

where to simplify the notation we write $f = f_{p_1}$. Using now (4.29) we get

$$f(r + s) = \frac{C_2}{f(r - s)} = \frac{C_2}{C_1}f(s - r) \tag{4.31}$$

Equation (4.31) shows that the function $\pi(s) := K^s f(s)$, where $K = (C_1/C_2)^{\frac{1}{2r}}$, is periodic with real period $2r$. Furthermore, since $p_1 > 0$, $\pi(s)$ is analytic in the strip of the width strictly larger than $2r$; thus it extends analytically to $\mathbb{C}$. By Lemma 4.1.8 this determines uniquely the Mellin transform of $|X|$. Namely,

$$E|X|^s = CK^s E|Z|^s.$$

Therefore in distribution we have the representation

$$X \cong KZ\chi, \tag{4.32}$$

where K is a constant, Z is normal N(0,1), and χ is a $\{0, 1\}$-valued independent of Z random variable such that $P(\chi = 1) = C$.

Clearly, the proof is concluded if $C = 0$ (X being degenerate normal). If $C \neq 0$ then by (4.32)

$$E|tX + uY|^p \tag{4.33}$$
$$= C(1 - C)^2(t^2 + u^2)^{p/2}E|Z|^p + C(1 - C)(|t|^p + |u|^p)E|Z|^p.$$

Therefore $C = 1$, which ends the proof. $\square$

The next result comes from [23] and uses stringent moment conditions; Braverman [16] gives examples which imply that the condition on zeros of the Mellin transform cannot be dropped.

Theorem 4.2.9 *Let X, Y be symmetric i. i. d. random variables such that*

$$E\exp(\lambda|X|^2) < \infty$$

for some $\lambda > 0$, and $E|X|^s \neq 0$ for all $s \in \mathbb{C}$ such that $\Re s > 0$. Suppose there is a constant C such that for all real a, b

$$E|aX + bY| = C(a^2 + b^2)^{1/2}.$$

Then X, Y are normal.

The rest of this section is devoted to the proof of Theorem 4.2.9.

The function $\phi(s) = E|X|^s$ is analytic in the half–plane $\Re s > 0$. Since $E|Z|^s = \pi^{-1/2}K^s\Gamma(\frac{s+1}{2})$, where $K = \pi^{1/2}E|Z| > 0$ and $\Gamma(.)$ is the Euler gamma function, therefore (4.21) means that $\phi(s) = \pi^{-1/2}K^s\alpha(s)\Gamma(\frac{s+1}{2})$, where $\alpha(s) := \pi^{1/2}K^{-s}\phi(s)/\Gamma(\frac{s+1}{2})$ is analytic in the half-plane $\Re s > 0, \alpha(\bar{s}) = \overline{\alpha(s)}$ and satisfies

$$\alpha(s)\alpha(1 - s) = 1 \text{ for } 0 < \Re s < 1. \tag{4.34}$$

We shall need the following estimate, in which without loss of generality we may assume $0 < \lambda K < 1$ (choose $\lambda > 0$ small enough).

Lemma 4.2.10 *There is a constant $C > 0$ such that $|\alpha(s)| \leq C|s|(\lambda K)^{-\Re s}$ for all s in the half-plane $\Re s \geq \frac{1}{2}$.*

Proof. Since $E\exp(\lambda^2|X|^2) < \infty$ for some $\lambda > 0$, therefore $P(|X| \geq t) \leq Ce^{-\lambda^2 t^2}$, where $C = E\exp(\lambda^2|X|^2)$, see Problem 1.4. This implies

$$|\phi(s)| \leq C_1|s|\lambda^{-\Re s}\Gamma(\frac{1}{2}\Re s), \Re s > 0. \tag{4.35}$$

In particular $|\alpha(s)| \leq C\exp(o(|s|^2))$, where $o(x)/x \to 0$ as $x \to \infty$.

Consider now function $u(s) = \alpha(s)(\lambda K)^s/s$, which is analytic in $\Re s > 0$. Clearly $|u(s)| \leq C\exp(o(|s|^2))$ as $|s| \to \infty$. Moreover $|u(\frac{1}{2} + it)| \leq const$ for all real t by (4.34); for all real x

$$|u(x)| = \pi^{1/2}x^{-1}\lambda^x\phi(x)/\Gamma(\frac{x+1}{2}) \leq C_1\Gamma(\frac{1}{2}x)/\Gamma(\frac{x+1}{2}) \leq \pi^{1/2}C,$$

by (4.35). Therefore by the Phragmén-Lindelöf principle, see, eg. [97, page 50 Theorem 22], applied twice to the angles $-\frac{1}{2}\pi \leq \arg s \leq 0$, and $0 \leq \arg s \leq \frac{1}{2}\pi$, the Lemma is proved. $\square$

By Lemma 4.2.5 Theorem 4.2.9 follows from the next result.

Lemma 4.2.11 *Suppose X is a symmetric random variable satisfying*

$$E\exp(\lambda^2|X|^2) < \infty$$

for some $\lambda > 0$, and

$$E|X|^s \neq 0$$

for all $s \in C$, such that $\Re s > 0$. Let Z be a centered normal random variable such that

$$E|X|^{1/2+it}E|X|^{1/2-it} = E|Z|^{1/2+it}E|Z|^{1/2-it} \tag{4.36}$$

for all $t \in \mathbb{R}$. Then X is normal.

Proof.

We shall use Lemma 4.2.10 to show that $\alpha(s) = C_1 C_2^s$ for some real $C_1, C_2 > 0$. It is clear that $\alpha(s) \neq 0$ if $\Re s > 0$. Therefore $\beta(s) = \log \alpha(s)$ is a well defined function which is analytic in the half-plane $\Re s > 0$. The function $v(s) := \Re(\beta(-is)) = \log|\alpha(-is)|$ is harmonic in the half-plane $\Im s > -\frac{1}{2}$ and $\limsup_{|s|\to\infty} v(s)/|s| < \infty$ by Lemma 4.2.10. Furthermore by (4.34) we have $v(t) = 0$ for real t. By the Nevanlina integral representation, see [97, page 233, Theorem 4]

$$v(x + iy) = \frac{y}{\pi}\int_{-\infty}^{\infty} \frac{v(t)}{(t-x)^2 + y^2}\,dt + ky$$

for some real constant k and for all real x, y with $y > 0$. This in particular implies that $\beta(y + \frac{1}{2}) = \Re(\beta(y + \frac{1}{2})) = v(-iy) = cy$. Thus by the uniqueness of analytic extension we get $\alpha(s) = C_1 C_2^s$ and hence

$$\phi(s) = \pi^{-1/2}K^s C_1 C_2^s \Gamma(\frac{s+1}{2}) \tag{4.37}$$

for some constants C_1, C_2 such that $C_1^2 C_2 = 1$ (the latter is the consequence of (4.34)). Formula (4.37) shows that the distribution of X is given by (4.32). To exclude the possibility that $P(X = 0) \neq 0$ it remains to verify that $C_1 = 1$. This again follows from (4.33). By Theorem 1.8.1, the proof is completed. $\square$

4.3 Infinite spherically symmetric sequences

In this section we present results that hold true for infinite sequences only and which might fail for finite sequences.

Definition 4.3.1 *An infinite sequence $X_1, X_2, \ldots$ is spherically symmetric if the finite sequence $X_1, X_2, \ldots, X_n$ is spherically symmetric for all n.*

The following provides considerably more information than Theorem 4.1.2.

Theorem 4.3.1 ([132]) *If an infinite sequence $\mathbf{X} = (X_1, X_2, \ldots)$ is spherically symmetric, then there is a sequence of independent identically distributed Gaussian random variables $\vec{\gamma} = (\gamma_1, \gamma_2, \ldots)$ and a non-negative random variable R independent of $\vec{\gamma}$ such that*

$$\mathbf{X} = R\vec{\gamma}.$$

This result is based on exchangeability.

Definition 4.3.2 *A sequence (X_k) of random variables is exchangeable, if the joint distribution of $X_{\sigma(1)}, X_{\sigma(2)}, \ldots, X_{\sigma(n)}$ is the same as the joint distribution of $X_1, X_2, \ldots, X_n$ for all $n \geq 1$ and for all permutations σ of $\{1, 2, \ldots, n\}$.*

Clearly, spherical symmetry implies exchangeability. The following beautiful theorem due to B. de Finetti [31] points out the role of exchangeability in characterizations as a substitute for independence; for more information and the references see [79].

Theorem 4.3.2 *Suppose that $X_1, X_2, \ldots$ is an infinite exchangeable sequence. Then there exist a σ-field $\mathcal{N}$ such that $X_1, X_2, \ldots$ are $\mathcal{N}$-conditionally i. i. d., that is*

$$P(X_1 < a_1, X_2 < a_2, \ldots, X_n < a_n | \mathcal{N})$$

$$= P(X_1 < a_1 | \mathcal{N}) P(X_1 < a_2 | \mathcal{N}) \ldots P(X_1 < a_n | \mathcal{N})$$

for all $a_1, \ldots, a_n \in \mathbb{R}$ and all $n \geq 1$.

Proof. Let $\mathcal{N}$ be the tail σ-field, ie.

$$\mathcal{N} = \bigcap_{k=1}^{\infty} \sigma(X_k, X_{k+1}, \ldots)$$

and put $\mathcal{N}_k = \sigma(X_k, X_{k+1}, \ldots)$. Fix bounded measurable functions f, g, h and denote

$$F_n = f(X_1, \ldots, X_n);$$

$$G_{n,m} = g(X_{n+1}, \ldots, X_{m+n});$$

$$H_{n,m,N} = h(X_{m+n+N+1}, X_{m+n+N+2}, \ldots),$$

where $n, m, N \geq 1$. Exchangeability implies that

$$E F_n G_{n,m} H_{n,m,N} = E F_n G_{n+r,m} H_{n,m,N}$$

for all $r \leq N$. Since $H_{n,m,N}$ is an arbitrary bounded $\mathcal{N}_{m+n+N+1}$-measurable function, this implies

$$E\{F_n G_{n,m} | \mathcal{N}_{m+n+N+1}\} = E\{F_n G_{n+r,m} | \mathcal{N}_{m+n+N+1}\}.$$

Passing to the limit as $N \to \infty$, see Theorem 1.4.3, this gives

$$E\{F_n G_{n,m} | \mathcal{N}\} = E\{F_n G_{n+r,m} | \mathcal{N}\}.$$

Therefore

$$E\{F_n G_{n,m} | \mathcal{N}\} = E\{G_{n+r,m} E\{F_n | \mathcal{N}_{n+r+1}\} | \mathcal{N}\}.$$

Since $E\{F_n|\mathcal{N}_{n+r+1}\}$ converges in L_1 to $E\{F_n|\mathcal{N}\}$ as $r \to \infty$, and since g is bounded,

$$E\{G_{n+r,m}E\{F_n|\mathcal{N}_{n+r+1}\}|\mathcal{N}\}$$

is arbitrarily close (in the L_1 norm) to

$$E\{G_{n+r,m}E\{F_n|\mathcal{N}\}|\mathcal{N}\} = E\{F_n|\mathcal{N}\}E\{G_{n+r,m}|\mathcal{N}\}$$

as $r \to \infty$. By exchangeability $E\{G_{n+r,m}|\mathcal{N}\} = E\{G_{n,m}|\mathcal{N}\}$ almost surely, which proves that

$$E\{F_nG_{n,m}|\mathcal{N}\} = E\{F_n|\mathcal{N}\}E\{G_{n,m}|\mathcal{N}\}.$$

Since f, g are arbitrary, this proves $\mathcal{N}$-conditional independence of the sequence. Using the exchangeability of the sequence once again, one can see that random variables $X_1, X_2, \ldots$ have the same $\mathcal{N}$-conditional distribution and thus the theorem is proved. $\square$

Proof of Theorem 4.3.1. Let $\mathcal{N}$ be the tail σ-field as defined in the proof of Theorem 4.3.2. By assumption, sequences

$$(X_1, X_2, \ldots),$$
$$(-X_1, X_2, \ldots),$$
$$(2^{-1/2}(X_1 + X_2), X_3, \ldots),$$
$$(2^{-1/2}(X_1 + X_2), 2^{-1/2}(X_1 - X_2), X_3, X_4, \ldots)$$

are all identically distributed and all have the same tail σ-field $\mathcal{N}$. Therefore, by Theorem 4.3.2 random variables X_1, X_2, are $\mathcal{N}$-conditionally independent and identically distributed; moreover, each variable has the symmetric $\mathcal{N}$-conditional distribution and $\mathcal{N}$-conditionally X_1 has the same distribution as $2^{-1/2}(X_1 + X_2)$. The rest of the argument repeats the proof of Theorem 3.1.1. Namely, consider conditional characteristic function $\phi(t) = E\{\exp(itX_1)|\mathcal{N}\}$. With probability one $\phi(1)$ is real by $\mathcal{N}$-conditional symmetry of distribution and $\phi(t) = (\phi(2^{-1/2}t))^2$. This implies

$$\phi(2^{-n/2}) = (\phi(1))^{1/2^n} \tag{4.38}$$

almost surely, $n = 0, 1, \ldots$. Since $\phi(2^{-n/2}) \to \phi(0) = 1$ with probability 1, we have $\phi(1) \neq 0$ almost surely. Therefore on a subset $\Omega_0 \subset \Omega$ of probability $P(\Omega_0) = 1$, we have $\phi(1) = \exp(-R^2)$, where $R^2 \geq 0$ is $\mathcal{N}$-measurable random variable. Applying[5] Corollary 2.3.4 for each fixed $\omega \in \Omega_0$ we get that $\phi(t) = \exp(-tR^2)$ for all real t.
$\square$

The next corollary shows how much simpler the theory of infinite sequences is, compare Theorem 4.1.4.

---

[5]Here we swept some dirt under the rug: the argument goes through, if one knows that except on a set of measure 0, $\phi(.)$ is a characteristic function. This requires using regular conditional distributions, see, eg. [9, Theorem 33.3.].

Corollary 4.3.3 *Let* $\mathbf{X} = (X_1, X_2, \ldots)$ *be an infinite spherically symmetric sequence such that* $E|X_k|^\alpha < \infty$ *for some real* $\alpha > 0$ *and all* $k = 1, 2, \ldots$ *Suppose that for some* $m \geq 1$

$$E\{\|(X_1, \ldots, X_m)\|^\alpha | (X_{m+1}, X_{m+2}, \ldots)\} = const. \tag{4.39}$$

Then $\mathbf{X}$ *is Gaussian.*

Proof. From Theorem 4.3.1 it follows that

$$E\{\|(X_1, \ldots, X_m)\|^\alpha | (X_{m+1}, X_{m+2}, \ldots)\}$$

$$= E\{R^\alpha \|(\gamma_1, \ldots, \gamma_m)\|^\alpha | (X_{m+1}, X_{m+2}, \ldots)\}.$$

However, R is measurable with respect to the tail σ-field, and hence it also is $\sigma(X_{m+1}, X_{m+2}, \ldots)$-measurable for all m. Therefore

$$E\{\|(X_1, \ldots, X_m)\|^\alpha | (X_{m+1}, X_{m+2}, \ldots)\}$$

$$= R^\alpha E\{\|(\gamma_1, \ldots, \gamma_m)\|^\alpha | R(\gamma_{m+1}, \gamma_{m+2}, \ldots)\}$$

$$= R^\alpha E\left\{ E\{\|(\gamma_1, \ldots, \gamma_m)\|^\alpha | R, (\gamma_{m+1}, \gamma_{m+2}, \ldots)\} | R(\gamma_{m+1}, \gamma_{m+2}, \ldots)\right\}.$$

Since R and $\vec{\gamma}$ are independent, we finally get

$$E\{\|(X_1, \ldots, X_m)\|^\alpha | (X_{m+1}, X_{m+2}, \ldots)\}$$

$$= R^\alpha E\{\|(\gamma_1, \ldots, \gamma_m)\|^\alpha | (\gamma_{m+1}, \gamma_{m+2}, \ldots)\} = C_\alpha R^\alpha.$$

Using now (4.39) we have $R = const$ almost surely and hence $\mathbf{X}$ is Gaussian. $\square$

The following corollary of Theorem 4.3.2 deals with exponential distributions as defined in Section 3.5. Diaconis & Freedman [35] have a dozen of de Finetti–style results, including this one.

Theorem 4.3.4 *If* $\mathbf{X} = (X_1, X_2, \ldots)$ *is an infinite sequence of non-negative random variables such that random variable* $\min\{X_1/a_1, \ldots, X_n/a_n\}$ *has the same distribution as* $(a_1 + \ldots + a_n)^{-1} X_1$ *for all* n *and all* $a_1, \ldots, a_n > 0$ *, then* $\mathbf{X} = \Lambda \vec{\epsilon}$*, where* Λ *and* $\vec{\epsilon}$ *are independent random variables and* $\vec{\epsilon} = (\epsilon_1, \epsilon_2, \ldots)$ *is a sequence of independent identically distributed exponential random variables.*

Sketch of the proof: Combine Theorem 3.4.1 with Theorem 4.3.2 to get the result for the pair X_1, X_2. Use the reasoning from the proof of Theorem 3.4.1 to get the representation for any finite sequence $X_1, \ldots, X_n$, see also Proposition 3.5.1.

4.4 Problems

Problem 4.1 *Prove the converse of (4.2). Namely, if $\phi(s)$ is the characteristic function of a one-dimensional random variable, then there is a spherically symmetric $(X_1, \ldots, X_n)$ such that $\phi(\|\mathbf{t}\|)$ is its characteristic function.*

Problem 4.2 *For centered bivariate normal r. v. X, Y with variances 1 and correlation coefficient ρ (see Example 2.2.1), show that $E\{|X|\,|Y|\} = \frac{2}{\pi}(\sqrt{1-\rho^2} + \rho \arcsin \rho)$.*

Problem 4.3 *Let X, Y be i. i. d. random variables with the probability density function defined by $f(x) = C|x|^{-3}\exp(-1/x^2)$, where C is a normalizing constant, and $x \in \mathbb{R}$. Show that for any choice of $a, b \in \mathbb{R}$ we have*

$$E|aX + bY| = K(a^2 + b^2)^{1/2},$$

where $K = E|X|$.

Problem 4.4 *Using the methods used in the proof of Theorem 4.2.1 for $p = 1$ prove the following.*

> **Theorem 4.4.1** *Let $X, Y, Z \geq 0$ be i. i. d. and integrable random variables. Suppose that there is a constant $C \neq 0$ such that $E\min\{X/a, Y/c, Z/c\} = C/(a + b + c)$ for all $a, b, c > 0$. Then X, Y, Z are exponential.*

Problem 4.5 (deterministic analogue of theorem 4.2.1) *Show that if X, Y are independent with the same distribution, and $E|aX + bY| = 0$ for some $a, b \neq 0$, then X, Y are non-random.*

Chapter 5

Independent linear forms

In this chapter the property of interest is the independence of linear forms in independent random variables. In Section 5.1 we give a characterization result that is both simple to state and to prove; it is nevertheless of considerable interest. Section 5.2 parallels Section 3.2. We use the characteristic property of the normal distribution to define *abstract* group-valued Gaussian random variables. In this broader context we again obtain the zero-one law; we also prove an important result about the existence of exponential moments. In Section 5.3 we return to characterizations, generalizing Theorem 5.1.1. We show that the stochastic independence of arbitrary two linear forms characterizes the normal distribution. We conclude the chapter with *abstract Gaussian* results when *all forces* are joined.

5.1   Bernstein's theorem

The following result due to Bernstein [8] characterizes normal distribution by the independence of the sum and the difference of two independent random variables. More general but also more difficult result is stated in Theorem 5.3.1 below. An early precursor is Narumi [114], who proves a variant of Problem 5.4.The elementary proof below is adapted from Feller [54, Chapter 3].

Theorem 5.1.1 *If X_1, X_2 are independent random variables such that $X_1 + X_2$ and $X_1 - X_2$ are independent, then X_1 and X_2 are normal.*

The next result is an elementary version of Theorem 2.5.2.

Lemma 5.1.2 *If X, Z are independent random variables such that Z and $X + Z$ are normal, then X is normal.*

Indeed, the characteristic function ϕ of random variable X satisfies

$$\phi(t) \exp(-(t - m)^2/\sigma^2) = \exp(-(t - M)^2/S^2)$$

for some constants m, M, σ, S. Therefore $\phi(t) = \exp(at^2 + bt + c)$, for some real constants a, b, c, and by Proposition 2.1.1, ϕ corresponds to the normal distribution.

Lemma 5.1.3 *If X, Z are independent random variables and Z is normal, then $X + Z$ has a non-vanishing probability density function which has derivatives of all orders.*

Proof. Assume for simplicity that Z is $N(0, 2^{-1/2})$. Consider $f(x) = E\exp(-(x - X)^2)$. Then $f(x) \neq 0$ for each x, and since each derivative $\frac{d^k}{dy^k}\exp(-(y - X)^2)$ is bounded uniformly in variables y, X, therefore $f(\cdot)$ has derivatives of all orders. It remains to observe that $\pi^{-1/2}f(\cdot)$ is the probability density function of $X + Z$. This is easily verified using the cumulative distribution function:

$$P(X + Z \leq t) = \pi^{-1/2} \int_{-\infty}^{\infty} \exp(-z^2) \int_{\Omega} I_{X \leq t-z}\, dP\, dz$$

$$= \pi^{-1/2} \int_{\Omega} \left\{ \int_{-\infty}^{\infty} \exp(-z^2) I_{z+X \leq t}\, dz \right\} dP$$

$$= \pi^{-1/2} \int_{\Omega} \left\{ \int_{-\infty}^{\infty} \exp(-(y - X)^2) I_{y \leq t}\, dy \right\} dP$$

$$= \pi^{-1/2} \int_{-\infty}^{t} E\exp(-(y - X)^2)\, dy.$$

$\square$

Proof of Theorem 5.1.1. Let Z_1, Z_2 be i. i. d. normal random variables, independent of X's. Then random variables $Y_k = X_k + Z_k, k = 1, 2$, satisfy the assumptions of the theorem, cf. Theorem 2.2.6. Moreover, by Lemma 5.1.3, each of Y_k's has a smooth non-zero probability density function $f_k(x), k = 1, 2$. The joint density of the pair $Y_1 + Y_2, Y_1 - Y_2$ is $\frac{1}{2}f_1(\frac{x+y}{2})f_2(\frac{x-y}{2})$ and by assumption it factors into the product of two functions, the first being the function of x, and the other being the function of y only. Therefore the logarithms $Q_k(x) := \log f_k(\frac{1}{2}x), k = 1, 2$, are twice differentiable and satisfy

$$Q_1(x + y) + Q_2(x - y) = a(x) + b(y) \tag{5.1}$$

for some twice differentiable functions a, b (actually $a = Q_1 + Q_2$). Taking the mixed second order derivative of (5.1) we obtain

$$Q_1''(x + y) = Q_2''(x - y). \tag{5.2}$$

Taking $x = y$ this shows that $Q_1''(x) = const.$ Similarly taking $x = -y$ in (5.2) we get that $Q_2''(x) = const.$ Therefore $Q_k(2x) = A_k + B_k x + C_k x^2$, and hence $f_k(x) = \exp(A_k + B_k x + C_k x^2), k = 1, 2$. As a probability density function, f_k has to be integrable, $k = 1, 2$. Thus $C_k < 0$, and then $A_k = -\frac{1}{2}\log(-2\pi C_k)$ is determined uniquely from the condition that $\int f_k(x)\, dx = 1$. Thus $f_k(x)$ is a normal density and Y_1, Y_2 are normal. By Lemma 5.1.2 the theorem is proved. $\square$

5.2   Gaussian distributions on groups

In this section we shall see that the conclusion of Theorem 5.1.1 is related to integrability just as the conclusion of Theorem 3.1.1 is related to the fact that the normal distribution is a limit distribution for sums of i. i. d. random variables, see Problem 3.3.

Let $\mathbf{G}$ be a group with a σ-field $\mathcal{F}$ such that group operation $\mathbf{x}, \mathbf{y} \mapsto \mathbf{x} + \mathbf{y}$, is a measurable transformation $(\mathbf{G} \times \mathbf{G}, \mathcal{F} \otimes \mathcal{F}) \to (\mathbf{G}, \mathcal{F})$. Let $(\Omega, \mathcal{M}, P)$ be a probability space. A measurable function $\mathbf{X} : (\Omega, \mathcal{M}) \to (\mathbf{G}, \mathcal{F})$, is called a $\mathbf{G}$-valued random variable and its distribution is called a probability measure on $\mathbf{G}$.

Example 5.2.1 *Let $\mathbf{G} = \mathbb{R}^d$ be the vector space of all real d-tuples with vector addition as the group operation and with the usual Borel σ-field $\mathcal{B}$. Then a $\mathbf{G}$-valued random variable determines a probability distribution on $\mathbb{R}^d$.*

Example 5.2.2 *Let $\mathbf{G} = S^1$ be the group of all complex numbers z such that $|z| = 1$ with multiplication as the group operation and with the usual Borel σ-field $\mathcal{F}$ generated by open sets. A distribution of $\mathbf{G}$-valued random variable is called a probability measure on S^1.*

Definition 5.2.1 *A $\mathbf{G}$-valued random variable $\mathbf{X}$ is $\mathcal{I}$-Gaussian (letter $\mathcal{I}$ stays here for* independence*) if random variables $\mathbf{X} + \mathbf{X}'$ and $\mathbf{X} - \mathbf{X}'$, where $\mathbf{X}'$ is an independent copy of $\mathbf{X}$, are independent.*

Clearly, any vector space is an Abelian group with vector addition as the group operation. In particular, we now have two possibly distinct notions of Gaussian vectors: the $\mathcal{E}$-Gaussian vectors introduced in Section 3.2 and the $\mathcal{I}$-Gaussian vectors introduced in this section. In general, it seems to be not known, when the two definitions coincide; [143] gives related examples that satisfy suitable versions of the *2-stability* condition (as in our definition of $\mathcal{E}$-Gaussian) without being $\mathcal{I}$-Gaussian.

Let us first check that at least in some simple situations both definitions give the same result.

Example 5.2.1 (continued) *If $\mathbf{G} = \mathbb{R}^d$ and $\mathbf{X}$ is an $\mathbb{R}^d$-valued $\mathcal{I}$-Gaussian random variable, then for all $a_1, a_2, \ldots, a_d \in \mathbb{R}$ the one-dimensional random variable $a_1 X(1) + a_2 X(2) + \ldots + a_d X(d)$ has the normal distribution. This means that $\mathbf{X}$ is a Gaussian vector in the usual sense, and in this case the definitions of $\mathcal{I}$-Gaussian and $\mathcal{E}$-Gaussian random variables coincide. Indeed, by Theorem 5.1.1, if $\mathcal{L} : \mathbf{G} \to \mathbb{R}$ is a measurable homomorphism, then the $\mathbb{R}$-valued random variable $X = \mathcal{L}(\mathbf{X})$ is normal.*

In many situations of interest the reasoning that we applied to $\mathbb{R}^d$ can be repeated and both the definitions are consistent with the usual interpretation of the Gaussian distribution. An important example is the vector space $C[0, 1]$ of all continuous functions on the unit interval.

To some extend, the notion of $\mathcal{I}$-Gaussian variable is more versatile. It has wider applicability because less algebraic structure is required. Also there is some flexibility in the choice of the linear forms; the particular linear combination $\mathbf{X} + \mathbf{X}'$ and $\mathbf{X} - \mathbf{X}'$ seems to be quite arbitrary, although it might be a bit simpler for algebraic manipulations, compare the proofs of Theorem 5.2.2 and Lemma 5.3.2 below. This is quite different from Section 3.2; it is known, see [73, Chapter 2] that even in the real case not every pair of linear forms could be used to define an $\mathcal{E}$-Gaussian random variable. Besides, $\mathcal{I}$-Gaussian variables satisfy the following variant of $\mathcal{E}$-condition. In analogy with Section 3.2, for any $\mathbf{G}$-valued random variable $\mathbf{X}$ we may say that $\mathbf{X}$ is $\mathcal{E}'$-Gaussian, if $2\mathbf{X}$ has the same distribution as $\mathbf{X}_1 + \mathbf{X}_2 + \mathbf{X}_3 + \mathbf{X}_4$, where $\mathbf{X}_1, \mathbf{X}_2, \mathbf{X}_3, \mathbf{X}_4$ are four independent copies of

X. Any symmetric $\mathcal{I}$-Gaussian random variable is always $\mathcal{E}'$-Gaussian in the above sense, compare Problem 5.1. This observation allows to repeat the proof of Theorem 3.2.1 in the $\mathcal{I}$-Gaussian case, proving the zero-one law. For simplicity, we chose to consider only random variables with values in a vector space V; notation $2^n\mathbf{x}$ makes sense also for groups – the reader may want to check what goes wrong with the argument below for non-Abelian groups.

Theorem 5.2.1 *If* $\mathbf{X}$ *is a* V*-valued* $\mathcal{I}$*-Gaussian random variable and* $\mathbb{L}$ *is a linear measurable subspace of* V*, then* $P(\mathbf{X} \in \mathbb{L})$ *is either 0, or 1.*

Indeed, let $\mathbf{X}_1, \ldots, \mathbf{X}_n, \ldots$ be independent copies of $\mathbf{X}$, taken to be also independent from $\mathbf{X}$. Recurrently we see that $\mathbf{X}_1 + \ldots + \mathbf{X}_{4^n}$ and $2^n\mathbf{X}$ have the same distribution for all $n \geq 1$. Since $\mathbb{L}$ is a linear subspace of V, we have $P(\mathbf{X}_1 + \ldots + \mathbf{X}_{4^n} \in \mathbb{L}) = P(\mathbf{X} \in \mathbb{L})$. Put $\mathbf{Z} = \mathbf{X}_1 + \mathbf{X}_2 + \mathbf{X}_3 + \mathbf{X}_4$. Since $\mathbf{X}_1 + \ldots + \mathbf{X}_{4^{n+1}}$ has the same distribution as $\mathbf{Z} + 2^n\mathbf{X}$, therefore $P(\mathbf{Z} + 2^n\mathbf{X} \in \mathbb{L}) = P(\mathbf{X} \in \mathbb{L})$ does not depend on n. As in the proof of Theorem 3.2.1, define events $A_n = \{\mathbf{Z} \notin \mathbb{L}\} \cap \{\mathbf{Z} + 2^n\mathbf{X} \in \mathbb{L}\}$. It is again easily verified that events $\{A_n\}_{n \geq 1}$ are disjoint; therefore $P(A_n) = P(\mathbf{X} \in \mathbb{L})P(\mathbf{X} \notin \mathbb{L}) = 0$. $\square$

The main result of this section, Theorem 5.2.2, needs additional notation. This notation is natural for linear spaces. Let $\mathbf{G}$ be a group with a translation invariant metric $d(\mathbf{x}, \mathbf{y})$, ie. suppose $d(\mathbf{x} + \mathbf{z}, \mathbf{y} + \mathbf{z}) = d(\mathbf{x}, \mathbf{y})$ for all $\mathbf{x}, \mathbf{y}, \mathbf{z} \in \mathbf{G}$. Such a metric $d(\cdot, \cdot)$ is uniquely defined by the function $x \mapsto D(\mathbf{x}) := d(\mathbf{x}, 0)$. Moreover, it is easy to see that $D(\mathbf{x})$ has the following properties: $D(\mathbf{x}) = D(-\mathbf{x})$ and $D(\mathbf{x} + \mathbf{y}) \leq D(\mathbf{x}) + D(\mathbf{y})$ for all $\mathbf{x}, \mathbf{y} \in \mathbf{G}$. Indeed, by translation invariance $D(-\mathbf{x}) = d(-\mathbf{x}, 0) = d(0, \mathbf{x}) = d(\mathbf{x}, 0)$ and $D(\mathbf{x} + \mathbf{y}) = d(\mathbf{x} + \mathbf{y}, 0) \leq d(\mathbf{x} + \mathbf{y}, \mathbf{y}) + d(\mathbf{y}, 0) = D(\mathbf{x}) + D(\mathbf{y})$.

Theorem 5.2.2 *Let* $\mathbf{G}$ *be a group with a measurable translation invariant metric* $d(., .)$*. If* $\mathbf{X}$ *is an* $\mathcal{I}$*-Gaussian* $\mathbf{G}$*-valued random variable, then* $E\exp \lambda d(\mathbf{X}, 0) < \infty$ *for some* $\lambda > 0$.

More information can be gained in concrete situations. To mention one such example of great importance, consider a $C[0, 1]$-valued $\mathcal{I}$-Gaussian random variable, ie. a Gaussian stochastic process with continuous trajectories. Theorem 5.2.2 says that

$$E\exp \lambda(\sup_{0 \leq t \leq 1} |X(t)|) < \infty$$

for some $\lambda > 0$. On the other hand, $C[0, 1]$ is a normed space and another (equivalent) definition applies; Theorem 5.4.1 below implies stronger integrability property

$$E\exp \lambda(\sup_{0 \leq t \leq 1} |X(t)|^2) < \infty$$

for some $\lambda > 0$. However, even the weaker conclusion of Theorem 5.2.2 implies that the real random variable $\sup_{0 \leq t \leq 1} |X(t)|$ has moment generating function and that all its moments are finite. Lemma 5.3.2 below is another application of the same line of reasoning.

Proof of Theorem 5.2.2. Consider a real function $N(x) := P(D(\mathbf{X}) \geq x)$, where as before $D(\mathbf{x}) := d(\mathbf{x}, 0)$. We shall show that there is x_0 such that

$$N(2x) \leq 8(N(x - x_0))^2 \tag{5.3}$$

for each $x \geq x_0$. By Corollary 1.3.7 this will end the proof.

Let $\mathbf{X}_1, \mathbf{X}_2$ be the independent copies of $\mathbf{X}$. Inequality (5.3) follows from the fact that event $\{D(\mathbf{X}_1) \geq 2x\}$ implies that either the event $\{D(\mathbf{X}_1) \geq 2x\} \cap \{D(\mathbf{X}_2) \geq 2x_0\}$, or the event $\{D(\mathbf{X}_1 + \mathbf{X}_2) \geq 2(x - x_0)\} \cap \{D(\mathbf{X}_1 - \mathbf{X}_2) \geq 2(x - x_0)\}$ occurs.

Indeed, let x_0 be such that $P(D(\mathbf{X}_2) \geq 2x_0) \leq \frac{1}{2}$. If $D(\mathbf{X}_1) \geq 2x$ and $D(\mathbf{X}_2) < 2x_0$ then $D(\mathbf{X}_1 \pm \mathbf{X}_2) \geq D(\mathbf{X}_1) - D(\mathbf{X}_2) \geq 2(x - x_0)$. Therefore using independence and the trivial bound $P(D(\mathbf{X}_1 + \mathbf{X}_2) \geq 2a) \leq P(D(\mathbf{X}_1) \geq a) + P(D(\mathbf{X}_2) \geq a)$, we obtain

$$P(D(\mathbf{X}_1) \geq 2x) \leq P(D(\mathbf{X}_1) \geq 2x)P(D(\mathbf{X}_2) \geq 2x_0)$$

$$+P(D(\mathbf{X}_1 + \mathbf{X}_2) \geq 2(x - x_0))P(D(\mathbf{X}_1 - \mathbf{X}_2) \geq 2(x - x_0))$$

$$\leq \frac{1}{2}N(2x) + 4N^2(x - x_0)$$

for each $x \geq x_0$. $\square$

More theory of Gaussian distributions on groups can be developed when more structure is available, although technical difficulties arise; for instance, the Cramer theorem (Theorem 2.5.2) fails on the torus, see Marcinkiewicz [107]. Series expansion questions (cf. Theorem 2.2.5 and the remark preceding Theorem 8.1.3) are studied in [24], see also references therein. One can also study Gaussian distributions on normed vector spaces. In Section 5.4 below we shall see to what extend this extra structure is helpful, for integrability question; there are deep questions specific to this situation, such as what are the properties of the distribution of the real r. v. $\|\mathbf{X}\|$; see [55]. Another research subject, entirely left out from this book, are Gaussian distributions on Lie groups; for more information see eg. [153]. Further information about abstract Gaussian random variables, can be found also in [27, 49, 51, 52].

5.3 Independence of linear forms

The next result generalizes Theorem 5.3.1 to more general linear forms of a given independent sequence $X_1, \ldots, X_n$. An even more general result that admits also zero coefficients in linear forms, was obtained independently by Darmois [30] and Skitovich [136]. Multidimensional variants of Theorem 5.3.1 are also known, see [73]. Banach space version of Theorem 5.3.1 was proved in [89].

Theorem 5.3.1 *If $X_1, \ldots, X_n$ is a sequence of independent random variables such that the linear forms $\sum_{k=1}^n a_k X_k$ and $\sum_{k=1}^n b_k X_k$ have all non-zero coefficients and are independent, then random variables X_k are normal for all $1 \leq k \leq n$.*

Our proof of Theorem 5.3.1 uses additional information about the existence of moments, which then allows us to use an argument from [104] (see also [75]). Notice that we don't allow for vanishing coefficients; the latter case is covered by [73, Theorem 3.1.1] but the proof is considerably more involved[1].

We need a suitable generalization of Theorem 5.2.2, which for simplicity we state here for real valued random variables only. The method of proof seems also to work in more general context under the assumption of independence of certain nonlinear statistics, compare [101, Section 5.3.], [73, Section 4.3] and Lemma 7.4.2 below.

Lemma 5.3.2 *Let* $a_1, \ldots, a_n, b_1, \ldots, b_n$ *be two sequences of non-zero real numbers. If* $X_1, \ldots, X_n$ *is a sequence of independent random variables such that two linear forms* $\sum_{k=1}^{n} a_k X_k$ *and* $\sum_{k=1}^{n} b_k X_k$ *are independent, then random variables* $X_k, k = 1, 2, \ldots, n$ *have finite moments of all orders.*

Proof. We shall repeat the idea from the proof of Theorem 5.2.2 with suitable technical modifications. Suppose that $0 < \epsilon \leq |a_k|, |b_k| \leq K < \infty$ for $k = 1, 2, \ldots, n$. For $x \geq 0$ denote $N(x) := \max_{j \leq n} P(|X_j| \geq x)$ and let $C = 2nK/\epsilon$. For $1 \leq j \leq n$ we have trivially

$$P(|X_j| \geq Cx) \leq P(|X_j| \geq Cx, |X_k| \leq x \, \forall k \neq j)$$

$$+ \sum_{k \neq j}^{n} P(|X_j| \geq x) P(|X_k| \geq x).$$

Notice that the event $A_j := \{|X_j| \geq Cx\} \cap \{|X_k| \leq x \, \forall k \neq j\}$ implies that both $|\sum_{k=1}^{n} a_k X_k| \geq nKx$ and $|\sum_{k=1}^{n} b_k X_k| \geq nKx$. Indeed,

$$|\sum_{k=1}^{n} a_k X_k| \geq |X_j||a_j| - \sum_{k, \, k \neq j} |a_k X_k| \geq (\epsilon C - nK)x = nKx$$

and the second inclusion follows analogously. By independence of the linear forms this shows that

$$P(|X_j| \geq Cx) \leq P(|\sum_{k=1}^{n} a_k X_k| \geq nKx) P(|\sum_{k=1}^{n} b_k X_k| \geq nKx)$$

$$+ \sum_{k \neq j}^{n} P(|X_j| \geq x) P(|X_k| \geq x).$$

Therefore $N(Cx) \leq P(|\sum_{k=1}^{n} a_k X_k| \geq nKx) P(|\sum_{k=1}^{n} b_k X_k| \geq nKx) + nN^2(x)$. Using the trivial bound

$$P(|\sum_{k=1}^{n} a_k X_k| \geq nKx) \leq nN(x),$$

we get

$$N(Cx) \leq 2n^2 N^2(x).$$

Corollary 1.3.3 now ends the proof. $\square$

Proof of Theorem 5.3.1. We shall begin with reducing the theorem to the case with more information about the coefficients of the linear forms. Namely, we shall reduce the proof to the case when all $a_k = 1$, and all b_k are different.

---

[1]The only essential use of non-vanishing coefficients is made in the proof of Lemma 5.3.2.

Since all a_k are non-zero, normality of X_k is equivalent to normality of $a_k X_k$; hence passing to $X_k' = a_k X_k$, we may assume that $a_k = 1, 1 \leq k \leq n$. Then, as the second step of the reduction, without loss of generality we may assume that all b_j's are different. Indeed, if, eg. $b_1 = b_2$, then substituting $X_1' = X_1 + X_2$ we get $(n-1)$ independent random variables $X_1', X_3, X_4, \ldots, X_n$ which still satisfy the assumptions of Theorem 5.3.1; and if we manage to prove that X_1' is normal, then by Theorem 2.5.2 the original random variables X_1, X_2 are normal, too.

The reduction argument allows without loss of generality to assume that $a_k = 1, 1 \leq k \leq n$ and $0 \neq b_1 \neq b_2 \neq \ldots \neq b_n$. In particular, the coefficients of linear forms satisfy the assumption of Lemma 5.3.2. Therefore random variables $X_1, \ldots, X_n$ have finite moments of all orders and linear forms $\sum_{k=1}^n X_k$ and $\sum_{k=1}^n b_k X_k$ are independent.

The joint characteristic function of $\sum_{k=1}^n X_k, \sum_{k=1}^n b_k X_k$ is

$$\phi(t, s) = \prod_{k=1}^n \phi_k(t + b_k s),$$

where ϕ_k is the characteristic function of random variable $X_k, k = 1, \ldots, n$. By independence of linear forms $\phi(t, s)$ factors

$$\phi(t, s) = \Psi_1(t)\Psi_2(s).$$

Hence

$$\prod_{k=1}^n \phi_k(t + b_k s) = \Psi_1(t)\Psi_2(s). \tag{5.4}$$

Passing to the logarithms $Q_k = \log \phi_k$ in a neighborhood of 0, from (5.4) we obtain

$$\sum_{k=1}^n Q_k(t + b_k s) = w_1(t) + w_2(s). \tag{5.5}$$

By Lemma 5.3.2 functions Q_k and w_j have derivatives of all orders, see Theorem 1.5.1. Consecutive differentiation of (5.5) with respect to variable s at $s = 0$ leads to the following system of equations

$$\begin{aligned}
\sum_{k=1}^n b_k Q_k'(t) &= w_2'(0), \\
\sum_{k=1}^n b_k^2 Q_k''(t) &= w_2''(0), \\
&\vdots \\
\sum_{k=1}^n b_k^n Q_k^{(n)}(t) &= w_2^{(n)}(0).
\end{aligned} \tag{5.6}$$

Differentiation with respect to t gives now

$$\begin{aligned}
\sum_{k=1}^n b_k Q_k^{(n)}(t) &= 0, \\
\sum_{k=1}^n b_k^2 Q_k^{(n)}(t) &= 0,
\end{aligned} \tag{5.7}$$

$$\vdots$$

$$\sum_{k=1}^{n} b_k^{n-1} Q_k^{(n)}(t) \;=\; 0,$$

$$\sum_{k=1}^{n} b_k^{n} Q_k^{(n)}(t) \;=\; const$$

(clearly, the last equation was not differentiated).

Equations (5.7) form a system of linear equations (5.7) for unknown values $Q_k^{(n)}(t)$, $1 \le k \le n$. Since all b_j are non-zero and different, therefore the determinant of the system is non-zero[2]. The unique solution $Q_k^{(n)}(t)$ of the system is $Q_k^{(n)}(t) = const_k$ and does not depend on t. This means that in a neighborhood of 0 each of the characteristic functions $\phi_k(\cdot)$ can be written as $\phi_k(t) = \exp(P_k(t))$, where P_k is a polynomial of at most n-th degree. Theorem 2.5.3 now concludes the proof. $\square$

Remark: Additional integrability information was used to solve equation (5.5). In general equation (5.5) has the same solution but the proof is more difficult, see [73, Section A.4.].

5.4 Strongly Gaussian vectors

Following Fernique, we give yet another definition of a Gaussian random variable.

Let V be a linear space and let X be an V-valued random variable. Denote by X' an independent copy of X.

Definition 5.4.1 X *is S-Gaussian (S stays here for* strong) *if for all real α random variables $\cos(\alpha)X' + \sin(\alpha)X$, and $\sin(\alpha)X' - \cos(\alpha)X$ are independent and have the same distribution as X.*

Clearly any S-Gaussian random vector is both $\mathcal{I}$-Gaussian and $\mathcal{E}$-Gaussian, which motivates the adjective "strong". Let us quickly show how Theorems 3.2.1 and 5.2.1 can be obtained for S-Gaussian vectors. The proofs follow Fernique [55].

Theorem 5.4.1 *If X is an V -valued S-Gaussian random variable and $\mathbb{L}$ is a linear measurable subspace of V, then $P(X \in \mathbb{L})$ is either equal to 0, or to 1.*

Proof. Let X, X' be independent copies of X. For each $0 < \alpha < \pi/2$, let $X_\alpha = \cos(\alpha)X + \sin(\alpha)X'$, and consider the event

$$A(\alpha) = \{\omega : X_\alpha(\omega) \in \mathbb{L}\} \cap \{X_{\pi/2-\alpha}(\omega) \notin \mathbb{L}\}.$$

Clearly $P(A(\alpha)) = P(X \in \mathbb{L})P(X \notin \mathbb{L})$. Moreover, it is easily seen that $\{A(\alpha)\}_{0<\alpha<\pi/2}$ are pairwise disjoint events. Indeed, if $A(\alpha) \cap A(\beta) \ne \emptyset$, then we would have vectors $\mathbf{v}, \mathbf{w}$ such that $\cos(\alpha)\mathbf{v} + \sin(\alpha)\mathbf{w} \in \mathbb{L}, \cos(\beta)\mathbf{v} + \sin(\beta)\mathbf{w} \in \mathbb{L}$, which for $\alpha \ne \beta$ implies that $\mathbf{v}, \mathbf{w} \in \mathbb{L}$. This contradicts $\cos(\pi/2 - \alpha)\mathbf{v} + \sin(\pi/2 - \alpha)\mathbf{w} \notin \mathbb{L}$. Therefore $P(A(\alpha)) = 0$ for each α and in particular $P(X \in \mathbb{L})P(X \notin \mathbb{L}) = 0$, which ends the proof. $\square$

---

[2]This is the Vandermonde determinant and it equals $b_1 \ldots b_n \prod_{j<i}(b_j - b_i)$.

The next result is taken from Fernique [56]. It strengthens considerably the conclusion of Theorem 3.2.2.

Theorem 5.4.2 *Let* V *be a normed linear space with the measurable norm* $\|\cdot\|$*. If* $\mathbf{X}$ *is an* S*-Gaussian* V*-valued random variable, then there is* $\epsilon > 0$ *such that* $E\exp(\epsilon\|\mathbf{X}\|^2) < \infty$*.*

Proof. As previously, let $N(x) := P(\|\mathbf{X}\| \geq x)$. Let $\mathbf{X}_1, \mathbf{X}_2$ be independent copies of $\mathbf{X}$. It follows from the definition that

$$\|\mathbf{X}_1\|, \|\mathbf{X}_2\|$$

and

$$2^{-1/2}\|\mathbf{X}_1 + \mathbf{X}_2\|, 2^{-1/2}\|\mathbf{X}_1 - \mathbf{X}_2\|$$

are two pairs of independent copies of $\|\mathbf{X}\|$. Therefore for any $0 \leq y \leq x$ we have the following estimate

$$N(x) = P(\|\mathbf{X}_1\| \geq x, \|\mathbf{X}_2\| \geq y) + P(\|\mathbf{X}_1\| \geq x, \|\mathbf{X}_2\| < y)$$

$$\leq N(x)N(y) + P(\|\mathbf{X}_1 + \mathbf{X}_2\| \geq x - y)P(\|\mathbf{X}_1 - \mathbf{X}_2\| \geq x - y).$$

Thus

$$N(x) \leq N(x)N(y) + N^2(2^{-1/2}(x - y)). \tag{5.8}$$

Take x_0 such that $N(x_0) \leq \frac{1}{2}$. Substituting $t = \sqrt{2}x$ in (5.8) we get

$$N(\sqrt{2}t) \leq 2N^2(t - t_0) \tag{5.9}$$

for each $t \geq t_0$. This is similar to, but more precise than (5.3). Corollary 1.3.6 ends the proof. $\square$

5.5 Joint distributions

Suppose $X_1, \ldots, X_n, n \geq 1$, are (possibly dependent) random variables such that the joint distribution of n linear forms $L_1, L_2, \ldots, L_n$ in variables $X_1, \ldots, X_n$ is given. Then, except in the degenerate cases, the joint distribution of $(L_1, L_2, \ldots, L_n)$ determines uniquely the joint distribution of $(X_1, \ldots, X_n)$. The point to be made here is that if $X_1, \ldots, X_n$ are independent, then even degenerate transformations provide a lot of information. This phenomenon is responsible for results in Chapters 3 and 5. More general results which have little to do with the Gaussian distribution are also known. For instance, if X_1, X_2, X_3 are independent, then the joint distribution $\mu(dx, dy)$ of the pair $X_1 - X_2, X_2 - X_3$ determines the distribution of X_1, X_2, X_3 up to a change of location, provided that the characteristic function of μ does not vanish, see [73, Addendum A.3]. This result was found independently by a number of authors, see [84, 119, 124]; for related results see also [86, 151]. Nonlinear functions were analyzed in [87] and the references therein.

5.6 Problems

Problem 5.1 *Let $X_1, X_2, \ldots$ and $Y_1, Y_2, \ldots$ be two sequences of i. i. d. copies of random variables X, Y respectively. Suppose X, Y have finite second moments and are such that $U = X + Y$ and $V = X - Y$ are independent. Observe that in distribution $X \cong X_1 = \frac{1}{2}(U + V) \cong \frac{1}{2}(X_1 + Y_1 + X_2 - Y_2)$, etc. Use this observation and the Central Limit Theorem to prove Theorem 5.1.1 under the additional assumption of finiteness of second moments.*

Problem 5.2 *Let X and Y be two independent identically distributed random variables such that $U = X + Y$ and $V = X - Y$ are also independent. Observe that $2X = U + V$ and hence the characteristic function $\phi(\cdot)$ of X satisfies equation $\phi(2t) = \phi(t)\phi(t)\phi(-t)$. Use this observation to prove Theorem 5.1.1 under the additional assumption of i. i. d.*

Problem 5.3 (Deterministic version of Theorem 5.1.1) *Suppose X, U, V are independent and $X + U, X + V$ are independent. Show that X is non-random.*

The next problem gives a one dimensional converse to Theorem 2.2.9.

Problem 5.4 (From [114]) *Let X, Y be (dependent) random variables such that for some number $\rho \neq 0, \pm 1$ both $X - \rho Y$ and Y are independent and also $Y - \rho X$ and X are independent. Show that (X, Y) has bivariate normal distribution.*

Chapter 6

Stability and weak stability

The stability problem is the question of to what extent the conclusion of a theorem is sensitive to small changes in the assumptions. Such description is, of course, vague until the questions of how to quantify the departures both from the conclusion and from the assumption are answered. The latter is to some extent arbitrary; in the characterization context, typically, stability reasoning depends on the ability to prove that small changes (measured with respect to some *measure* of smallness) in assumptions of a given characterization theorem result in small departures (measured with respect to one of the distances of distributions) from the normal distribution.

Below we present only one stability result; more about stability of characterizations can be found in [73, Chapter 9], see also [102]. In Section 6.2 we also give two results that establish what one may call *weak stability*. Namely, we establish that *moderate* changes in assumptions still preserve some properties of the normal distribution. Theorem 6.2.2 below is the only result of this chapter used later on.

6.1   Coefficients of dependence

In this section we introduce a class of measures of departure from independence, which we shall call *coefficients of dependence*. There is no natural *measure of dependence* between random variables; those defined below have been used to define strong mixing conditions in limit theorems; for the latter the reader is referred to [65]; see also [10, Chapter 4].

To make the definition look less arbitrary, at first we consider an infinite parametric family of measures of dependence. For a pair of σ-fields $\mathcal{F}, \mathcal{G}$ let

$$\alpha_{r,s}(\mathcal{F}, \mathcal{G}) = \sup\{\frac{|P(A \cap B) - P(A)P(B)|}{P(A)^r P(B)^s} : A \in \mathcal{F}, B \in \mathcal{G} \text{ non-trivial}\}$$

with the range of parameters $0 \leq r \leq 1, 0 \leq s \leq 1, r + s \leq 1$. Clearly, $\alpha_{r,s}$ is a number between 0 and 1. It is obvious that $\alpha_{r,s} = 0$ if and only if the σ-fields $\mathcal{F}, \mathcal{G}$ are independent. Therefore one could use each of the coefficients $\alpha_{r,s}$ as a measure of departure from independence.

Fortunately, among the infinite number of coefficients of dependence thus introduced, there are just four really distinct, namely $\alpha_{0,0}, \alpha_{0,1}, \alpha_{1,0}$, and $\alpha_{1/2,1/2}$. By this we mean that the convergence to zero of $\alpha_{r,s}$ (when the σ-fields $\mathcal{F}, \mathcal{G}$ vary) is equivalent to the convergence to 0 of one of the above four coefficients. And since $\alpha_{0,1}$ and $\alpha_{1,0}$ are mirror images of each other, we are actually left with three coefficients only.

The formal statement of this equivalence takes the form of the following inequalities.

Proposition 6.1.1 *If $r + s < 1$, then $\alpha_{r,s} \le (\alpha_{0,0})^{1-r-s}$.*
If $r + s = 1$ and $0 < r \le \frac{1}{2} \le s < 1$, then $\alpha_{r,s} \le (\alpha_{1/2,1/2})^{2r}$.

Proof. The first inequality follows from the fact that

$$\frac{|P(A \cap B) - P(A)P(B)|}{P(A)^r P(B)^s}$$

$$= |P(A \cap B) - P(A)P(B)|^{1-r-s}|P(B|A) - P(B)|^r|P(A|B) - P(A)|^s$$

$$\le |P(A \cap B) - P(A)P(B)|^{1-r-s}.$$

The second one is a consequence of

$$\frac{|P(A \cap B) - P(A)P(B)|}{P(A)^r P(B)^s}$$

$$= \left(\frac{|P(A \cap B) - P(A)P(B)|}{P(A)^{1/2}P(B)^{1/2}}\right)^{2r} |P(A|B) - P(A)|^{s-r} \le (\alpha_{1/2,1/2})^{2r}$$

$\square$

Coefficients $\alpha_{0,0}$ and $\alpha_{0,1}, \alpha_{1,0}$ are the basis for the definition of classes of stationary sequences called in the limit theorems literature *strong-mixing* and *uniform strong mixing* (called also *ϕ-mixing*); $\alpha_{1/2,1/2}$ is equivalent to the *maximal correlation* coefficient (6.3), which is the basis of the so called *ρ-mixing* condition. Monograph [39] gives recent exposition and relevant references; see also [42, pp. 380–385].

There is also a whole continuous spectrum of non-equivalent coefficients $\alpha_{r,s}$ when $r + s > 1$. As those coefficients may attain value ∞, they are less frequently used; one notable exception is $\alpha_{1,1}$, which is the basis of the so called *ψ-mixing* condition and occurs occasionally in the assumptions of some limit theorems. Condition equivalent to $\alpha_{1,1} < \infty$ and conditions related to $\alpha_{r,s}$ with $r+s > 1$ are also employed in large deviation theorems, see [34, condition (U) and Chapter 5].

The following bounds[1] for the covariances between random variables in $L_p(\mathcal{F})$ and in $L_q(\mathcal{F})$ will be used later on.

Proposition 6.1.2 *If X is $\mathcal{F}$-measurable with p-th moment finite $(1 \le p \le \infty)$ and Y is $\mathcal{G}$-measurable with q-th moment finite $(1 \le q \le \infty)$ and $1/p + 1/q \le 1$, then*

$$|EXY - EXEY| \tag{6.1}$$

$$\le 4(\alpha_{0,0})^{1-1/p-1/q}(\alpha_{1,0})^{1/p}(\alpha_{0,1})^{1/q}\|X\|_p\|Y\|_q$$

where $\|X\|_p = (E|X|^p)^{1/p}$ if $p < \infty$ and $\|X\|_\infty = \ \text{ess sup}|X|$.

---

[1] Similar results are also known for $\alpha_{0,0}$ and $\alpha_{1/2,1/2}$. The latter is more difficult and is due to R. Bradley, see [13, Theorem 2.2] and the references therein.

Proof. We shall prove the result for $p = 1, q = \infty$ and $p = q = \infty$ only; these are the only cases we shall actually need; for the general case, see eg. [46, page 347 Corollary 2.5] or [65].

Let $M = \text{ess sup}|Y|$. Switching the order of integration (ie. by Fubini's theorem) we get, see Problem 1.1,

$$|EXY - EXEY|$$
$$= \left| \int_{-\infty}^{\infty} \int_{-M}^{M} \left(P(X \geq t, Y \geq s) - P(X \geq t)P(Y \geq s) \right) dt\, ds \right|$$
$$\leq \int_{-\infty}^{\infty} \int_{-M}^{M} |P(X \geq t, Y \geq s) - P(X \geq t)P(Y \geq s)|\, dt\, ds. \tag{6.2}$$

Since $|P(X \geq t, Y \geq s) - P(X \geq t)P(Y \geq s)| \leq \alpha_{1,0}P(X \geq t)$ (which is good for positive t) and $|P(X \geq t, Y \geq s) - P(X \geq t)P(Y \geq s)| = |P(X < t, Y \geq s) - P(X < t)P(Y \geq s)| \leq \alpha_{1,0}P(X \leq t)$ (which works well for negative t), inequality (6.2) implies

$$|EXY - EXEY| \leq \alpha_{1,0} \int_0^{\infty} \int_{-M}^{M} P(X \geq t)\, dt\, ds$$

$$+\alpha_{1,0} \int_0^{\infty} \int_{-M}^{M} P(X \leq -t)\, dt\, ds = 2\alpha_{1,0}E|X|\, \|Y\|_{\infty}.$$

Similar argument using $|P(X \geq t, Y \geq s) - P(X \geq t)P(Y \geq s)| \leq \alpha_{0,0}$ gives

$$|EXY - EXEY| \leq 4\alpha_{0,0}\|X\|_{\infty}\|Y\|_{\infty}.$$

$\square$

6.1.1 Normal case

Here we review without proofs the relations between the dependence coefficients in the multivariate normal case. Ideas behind the proofs can be found in the solutions to the Problems 6.2, 6.4, and 6.5.

The first result points points out that the coefficients $\alpha_{0,1}$ and $\alpha_{1,0}$ are of little interest in the normal case.

Theorem 6.1.3 *Suppose* $(\mathbf{X}, \mathbf{Y}) \in \mathbb{R}^{d_1 + d_2}$ *are jointly normal and* $\alpha_{0,1}(\mathbf{X}, \mathbf{Y}) < 1$. *Then* $\mathbf{X}, \mathbf{Y}$ *are independent.*

Denote by ρ the *maximal correlation coefficient*

$$\rho = \sup\{corr(f(\mathbf{X})g(\mathbf{Y})) : f(\mathbf{X}), g(\mathbf{Y}) \in L_2\}. \tag{6.3}$$

The following estimate due to Kolmogorov & Rozanov [83] shows that in the normal case the maximal correlation coefficient (6.3) can be estimated by $\alpha_{0,0}$. In particular, in the normal case we have

$$\alpha_{1/2,1/2} \leq 2\pi\alpha_{0,0}.$$

Theorem 6.1.4 *Suppose* $\mathbf{X}, \mathbf{Y} \in \mathbb{R}^{d_1 + d_2}$ *are jointly normal. Then*

$$corr(f(\mathbf{X}), g(\mathbf{Y})) \leq 2\pi\alpha_{0,0}(\mathbf{X}, \mathbf{Y})$$

for all square integrable f, g.

The next inequality is known as the so called *Nelson's hypercontractive estimate* [116] and is of importance in mathematical physics. It is also known in general that inequality (6.4) implies a bound for maximal correlation, see [34, Lemma 5.5.11].

Theorem 6.1.5 *Suppose* $(\mathbf{X}, \mathbf{Y}) \in \mathbb{R}^{d_1+d_2}$ *are jointly normal. Then*

$$Ef(\mathbf{X})g(\mathbf{Y}) \leq \|f(X)\|_p \|g(Y)\|_p \tag{6.4}$$

for all p-integrable f, g, provided $p \geq 1 + \rho$, where ρ is the maximal correlation coefficient (6.3).

6.2 Weak stability

A weak version of the stability problem may be described as allowing relatively large departures from the assumptions of a given theorem. In return, only a selected part of the conclusion is to be preserved. In this section the part of the characterization conclusion that we want to preserve is integrability. This problem is of its own interest. Integrability results are often useful as a first step in some proofs, see the proof of Theorem 5.3.1, or the proof of Theorem 7.5.1 below.

As a simple example of weak stability we first consider Theorem 5.1.1, which says that for independent r. v. X, Y we have $\alpha_{1,0}(X + Y, X - Y) = 0$ only in the normal case. We shall show that if the coefficient of dependence $\alpha_{1,0}(X + Y, X - Y)$ is small, then the distribution of X still has some finite moments. The method of proof is an adaptation of the proof of Theorem 5.2.2.

Proposition 6.2.1 *Suppose X, Y are independent random variables such that random variables $X + Y$ and $X - Y$ satisfy $\alpha_{1,0}(X + Y, X - Y) < \frac{1}{2}$. Then X and Y have finite moments $E|X|^\beta < \infty$ for $\beta < -\log_2(2\alpha_{1,0})$.*

Proof. Let $N(x) = \max\{P(|X| \geq x), P(|Y| \geq x)\}$. Put $\alpha = \alpha_{1,0}$. We shall show that for each $\rho > 2\alpha$, there is $x_0 > 0$ such that

$$N(2x) \leq \rho N(x - x_0) \tag{6.5}$$

for all $x \geq x_0$.

Inequality (6.5) follows from the fact that the event $\{|X| \geq 2x\}$ implies that either $\{|X| \geq 2x\} \cap \{|Y| \geq 2y\}$ or $\{|X + Y| \geq 2(x - y)\} \cap \{|X - Y| \geq 2(x - y)\}$ holds (make a picture). Therefore, using the independence of X, Y, the definition of $\alpha = \alpha_{1,0}(X + Y, X - Y)$ and trivial bound $P(|X + Y| \geq a) \leq P(|X| \geq \frac{1}{2}a) + P(|Y| \geq \frac{1}{2}a)$ we obtain

$$P(|X| \geq 2x) \leq P(|X| \geq 2x)P(|Y| \geq 2y)$$

$$+P(|X + Y| \geq 2(x - y))(\alpha + P(|X - Y| \geq 2(x - y)))$$

$$\leq N(2x)N(2y) + 2\alpha N(x - y) + 4N^2(x - y).$$

For any $\epsilon > 0$ pick y so that $N(2y) \leq \epsilon/(1 + \epsilon)$. This gives $N(2x) \leq (1 + \epsilon)2\alpha N(x - y) + 4(1 + \epsilon)N^2(x - y)$ for all $x > y$. Now pick $x_0 \geq y$ such that $N(x - y) \leq \epsilon\alpha/(1 + \epsilon)$ for all $x > y$. Then

$$N(2x) \leq 2(1 + 2\epsilon)\alpha N(x - y) \leq 2(1 + 3\epsilon)\alpha N(x - x_0)$$

for all $x \geq x_0$. Since $\epsilon > 0$ is arbitrary, this ends the proof of (6.5).

By Theorem 1.3.1 inequality (6.5) concludes the proof, eg. by formula (1.2).

$\square$

In Chapter 7 we shall consider assumptions about conditional moments. In Section 7.5 we need the integrability result which we state below. The assumptions are motivated by the fact that a pair X, Y with the bivariate normal distribution has linear regressions $E\{X|Y\} = a_0 + a_1 Y$ and $E\{Y|X\} = b_0 + b_1 X$, see (2.8); moreover, since $X - (a_0 + a_1 Y)$ and Y are independent (and similarly $Y - (b_0 + b_1 X)$ and X are independent), see Theorem 2.2.9, therefore the conditional variances $Var(X|Y)$ and $Var(Y|X)$ are non-random. These two properties do not characterize the normal distribution, see Problem 7.7. However, the assumption that regressions are linear and conditional variances are constant might be considered as the departure from the assumptions of Theorem 5.1.1 on the one hand and from the assumptions of Theorem 7.5.1 on the other. The following somehow surprising fact comes from [20]. For similar implications see also [19] and [22, Theorem 2.2].

Theorem 6.2.2 *Let X, Y be random variables with finite second moments and suppose that*

$$E\{|X - (a_0 + a_1 Y)|^2 | Y\} \leq const \tag{6.6}$$

and

$$E\{|Y - (b_0 + b_1 X)|^2 | X\} \leq const \tag{6.7}$$

for some real numbers a_0, a_1, b_0, b_1 such that $a_1 b_1 \neq 0, 1, -1$. Then X, Y have finite moments of all orders.

In the proof we use the conditional version of Chebyshev's inequality stated as Problem 1.9.

Lemma 6.2.3 *If $\mathcal{F}$ is a σ-field and $E|X| < \infty$, then*

$$P(|X| > t|\mathcal{F}) \leq E\{|X| \, |\mathcal{F}\}/t$$

almost surely.

Proof. Fix $t > 0$ and let $A \in \mathcal{F}$. By the definition of the conditional expectation

$$\int_A P(|X| > t|\mathcal{F}) \, dP = E\{I_A I_{|X|>t}\} \leq E\{|X|/t I_A I_{|X|>t}\} \leq t^{-1} E\{|X|I_A\}.$$

This end the proof by Lemma 1.4.2. $\square$

Proof of Theorem 6.2.2. First let us observe that without losing generality we may assume $a_0 = b_0 = 0$. Indeed, by triangle inequality $(E\{|X - a_1 Y|^2|Y\})^{1/2} \leq |a_0| + (E\{|X - (a_0 + a_1 Y)|^2|Y\})^{1/2} \leq const$, and the analogous bound takes care of (6.7). Furthermore, by passing to $-X$ or $-Y$ if necessary, we may assume $a = a_1 > 0$ and $b = b_1 > 0$. Let

$N(x) = P(|X| \geq x) + P(|Y| \geq x)$. We shall show that there are constants $K, C > 0$ such that

$$N(Kx) \leq CN(x)/x^2. \tag{6.8}$$

This will end the proof by Corollary 1.3.4.

To prove (6.8) we shall proceed as in the proof of Theorem 5.2.2. Namely, the event $\{|X| \geq Kx\}$, where $x > 0$ is fixed and K will be chosen later, can be decomposed into the sum of two disjoint events $\{|X| \geq Kx\} \cap \{|Y| \geq x\}$ and $\{|X| \geq Kx\} \cap \{|Y| < x\}$. Therefore trivially we have

$$P(|X| \geq Kx) \leq P(|X| \geq x, |Y| \geq x) \tag{6.9}$$
$$+P(|X| \geq Kx, |Y| < x) \qquad = P_1 + P_2 \text{ (say) .}$$

For K large enough the second term on the right hand side of (6.9) can be estimated by conditional Chebyshev's inequality from Lemma 6.2.3. Using trivial estimate $|Y - bX| \geq b|X| - |Y|$ we get

$$P_2 \leq P(|Y - bX| \geq (Kb - 1)x, |X| \geq Kx) \tag{6.10}$$
$$= \int_{|X| \geq Kx} P(|Y - bX| \geq (Kb - 1)x | X) \, dP \leq const\, N(Kx)/x^2.$$

To estimate P_1 in (6.9), observe that the event $\{|X| \geq x\}$ implies that either $|X - aY| \geq Cx$, or $|Y - bX| \geq Cx$, where $C = |1 - ab|/(1 + a)$. Indeed, suppose both are not true, ie. $|Y - bX| < Cx$ and $|X - aY| < Cx$. Then we obtain trivially

$$|1 - ab||X| = |X - abX| \leq |X - aY| + a|Y - bX| < C(1 + a)x.$$

By our choice of C, this contradicts $|X| \geq x$.

Using the above observation and conditional Chebyshev's inequality we obtain

$$P_1 \leq P(|X - aY| \geq Cx, |Y| \geq x)$$

$$+P(|Y - bX| \geq Cx, |X| \geq x) \leq C_1 N(x)/x^2.$$

This, together with (6.9) and (6.10) implies $P(|X| \geq Kx) \leq CN(x)/x^2$ for any $K > 1/b$ with constant C depending on K but not on x. Similarly $P(|Y| \geq Kx) \leq CN(x)/x^2$ for any $K > 1/a$, which proves (6.8). $\square$

6.3   Stability

In this section we shall use the coefficient $\alpha_{0,0}$ to analyze the stability of a variant[2] of Theorem 5.1.1 which is based on the approach sketched in Problem 5.2.

Theorem 6.3.1 *Suppose X, Y are i. i. d. with the cumulative distribution function $F(\cdot)$. Assume that $EX = 0, EX^2 = 1$ and $E|X|^3 = K < \infty$ and let $\Phi(\cdot)$ denote the cumulative distribution function of the standard normal distribution. If $\alpha_{0,0}(X + Y; X - Y) < \epsilon$, then*

$$\sup_x |F(x) - \Phi(x)| \leq C(K)\epsilon^{1/3}. \tag{6.11}$$

---

[2]Compare [112]. The proof below is taken from [73, section 9.2].

The following corollary is a consequence of Theorem 6.3.1 and Proposition 6.2.1.

Corollary 6.3.2 *Suppose X, Y are i. i. d. with the cumulative distribution function $F(\cdot)$. Assume that $EX = 0, EX^2 = 1$. If $\alpha_{1,0}(X + Y; X - Y) < \epsilon$, then there is $C < \infty$ such that (6.11) holds.*

Indeed, by Proposition 6.2.1 the third moment exists if $\epsilon < e^{-3/2}$; choosing large enough C inequality (6.11) holds true trivially for $\epsilon \geq e^{-3/2}$.

The next lemma gives the estimate of the left hand side of (6.11) in terms of characteristic functions. Inequality (6.12) is called *smoothing inequality* – a name well motivated by the method of proof; it is due to Esseen [45].

Lemma 6.3.3 *Suppose F, G are cumulative distribution functions with the characteristic functions ϕ, ψ respectively. If G is differentiable, then for all $T > 0$*

$$\sup_x |F(x) - G(x)| \leq \frac{1}{\pi} \int_{-T}^{T} |\phi(t) - \psi(t)| \, dt/t + \frac{12}{\pi T} \sup_x |G'(x)|. \qquad (6.12)$$

Proof. By the approximation argument, it suffices to prove (6.12) for F, G differentiable and with integrable characteristic functions only. Indeed, one can approximate F uniformly by the cumulative distribution functions F_δ, obtained by convoluting F with the normal $N(0, \delta)$ distribution, compare Lemma 5.1.3. The approximation, clearly, does not affect (6.12). That is, if (6.12) holds true for the approximants, then it holds true for the actual cdf's as well.

Let f, g be the densities of F and G respectively. The inversion formula for characteristic functions gives

$$f(x) = \frac{1}{2\pi} \int_{-\infty}^{\infty} e^{-itx} \phi(t) \, dt,$$

$$g(x) = \frac{1}{2\pi} \int_{-\infty}^{\infty} e^{-itx} \psi(t) \, dt.$$

From this we obtain

$$F(x) - G(x) = \frac{i}{2\pi} \int_{-\infty}^{\infty} e^{-itx} \frac{\phi(t) - \psi(t)}{t} \, dt.$$

The latter formula can be checked, for instance, by verifying that both sides have the same derivative, so that they may differ by a constant only. The constant has to be 0, because the left hand side has limit 0 at ∞ (a property of cdf) and the right hand side has limit 0 at ∞ (eg. because we convoluted with the normal distribution while doing our approximation step; another way of seeing what is the asymptotic at ∞ of the right hand side is to use the Riemann-Lebesgue theorem, see eg. [9, p. 354 Theorem 26.1]).

This clearly implies

$$\sup_x |F(x) - G(x)| \leq \frac{1}{2\pi} \int_{-\infty}^{\infty} |\phi(t) - \psi(t)| \, dt/t. \qquad (6.13)$$

This inequality, while resembling (6.12), is not good enough; it is not preserved by our approximation procedure, and the right hand side is useless when the density of F doesn't exist. Nevertheless (6.13) would do, if one only knew that the characteristic functions vanish outside of a finite interval. To achieve this, one needs to consider one more convolution approximation, this time we shall use density $h_T(x) = \frac{1}{\pi T}\frac{1-\cos(Tx)}{x^2}$. We shall need the fact that the characteristic function $\eta_T(t)$ of $h_T(x)$ vanishes for $|t| \geq T$ (and we shall not need the explicit formula $\eta_T(t) = 1 - |t|/T$ for $|t| \leq T$, cf. Example 1.5.1). Denote by F_T and G_T the cumulative distribution functions corresponding to convolutions $f \star h_T$ and $g \star h_T$ respectively. The corresponding characteristic functions are $\phi(t)\eta_T(t)$ and $\psi(t)\eta_T(t)$ respectively and both vanish for $|t| \geq T$. Therefore, inequality (6.13) applied to F_T and G_T gives

$$\sup_x |F_T(x) - G_T(x)| \qquad\qquad (6.14)$$

$$\leq \tfrac{1}{2\pi} \int_{-T}^{T} |(\phi(t) - \psi(t))\eta_T(t)|\, dt/t \ \leq \frac{1}{2\pi}\int_{-T}^{T} |\phi(t) - \psi(t)|\, dt/t.$$

It remains to verify that $\sup_x |F_T(x) - G_T(x)|$ does not differ too much from $\sup_x |F(x) - G(x)|$. Namely, we shall show that

$$\sup_x |F(x) - G(x)| \leq 2\sup_x |F_T(x) - G_T(x)| + \frac{12}{\pi T}\sup_x |G'(x)|, \qquad (6.15)$$

which together with (6.14) will end the proof of (6.12). To verify (6.15), put $M = \sup_x |G'(x)|$ and pick x_0 such that

$$\sup_x |F(x) - G(x)| = |F(x_0) - G(x_0)|.$$

Such x_0 can be found, because F and G are continuous and $F(x) - G(x)$ vanishes as $x \to \pm\infty$. Suppose $\sup_x |F(x) - G(x)| = G(x_0) - F(x_0)$. (The other case: $\sup_x |F(x) - G(x)| = F(x_0) - G(x_0)$ is handled similarly, and is done explicitly in [54, XVI. §3]). Since F is non-decreasing, and the rate of growth of G is bounded by M, for all $s \geq 0$ we get

$$G(x_0 - s) - F(x_0 - s) \geq G(x_0) - F(x_0) - sM.$$

Now put $a = \frac{G(x_0)-F(x_0)}{2M}, t = x_0 + a, x = a - s$. Then for all $|x| \leq a$ we get

$$G(t - x) - F(t - x) \geq \frac{1}{2}(G(x_0) - F(x_0)) + Mx. \qquad (6.16)$$

Notice that

$$G_T(t) - F_T(t) = \frac{1}{\pi T}\int_{-\infty}^{\infty} (F(t - x) - G(t - x))(1 - \cos Tx)x^{-2}\, dx$$

$$\geq \frac{1}{\pi T}\int_{-a}^{a} (F(t - x) - G(t - x))(1 - \cos Tx)x^{-2}\, dx$$

$$- \sup_x |F(x) - G(x)|\frac{2}{\pi T}\int_{a}^{\infty} y^{-2}\, dy.$$

Clearly,

$$\sup_x |F(x) - G(x)|\frac{2}{\pi T}\int_{a}^{\infty} y^{-2}\, dy = (G(x_0) - F(x_0))\frac{2}{\pi T}a^{-1} = 4M/(\pi T)$$

by our choice of a. On the other hand (6.16) gives

$$\frac{1}{\pi T}\int_{-a}^{a}(F(t-x)-G(t-x))(1-\cos Tx)x^{-2}\,dx$$

$$\geq \frac{1}{\pi T}\int_{-a}^{a}Mx(1-\cos Tx)x^{-2}\,dx$$

$$+\frac{1}{2}(G(x_0)-F(x_0))(1-\frac{2}{\pi T}\int_{a}^{\infty}y^{-2}\,dy)$$

$$=\frac{1}{2}(G(x_0)-F(x_0))-2M/(\pi T);$$

here we used the fact that the first integral vanishes by symmetry. Therefore $G(x_0) - F(x_0) \leq 2(G_T(x_0+a) - F_T(x_0+a)) + 12M/(\pi T)$, which clearly implies (6.15). $\square$

Proof of Theorem 6.3.1. Clearly only small $\epsilon > 0$ are of interest. Throughout the proof C will denote a constant depending on K only, not always the same at each occurrence. Let $\phi(.)$ be the characteristic function of X. We have $E\exp it(X+Y)\exp it(X-Y) = \phi(2t)$ and $E\exp it(X+Y)E\exp it(X-Y) = (\phi(t))^3\phi(-t)$. Therefore by a complex valued variant of (6.1) with $p = q = \infty$, see Problem 6.1, we have

$$|\phi(2t) - (\phi(t))^3\phi(-t)| \leq 16\epsilon. \tag{6.17}$$

We shall use (6.12) with $T = \epsilon^{-1/3}$ to show that (6.17) implies (6.11). To this end we need only to establish that for some $C > 0$

$$\frac{1}{\pi T}\int_{-T}^{T}|\phi(t) - e^{-\frac{1}{2}t^2}|/t\,dt \leq C\epsilon^{1/3}. \tag{6.18}$$

Put $h(t) = \phi(t) - e^{-\frac{1}{2}t^2}$. Since $EX = 0, EX^2 = 1$ and $E|X|^3 < \infty$, we can choose $\epsilon > 0$ small enough so that

$$|h(t)| \leq C_0|t|^3 \tag{6.19}$$

for all $|t| \leq \epsilon^{1/3}$. From (6.17) we see that

$$|h(2t)| = |\phi(2t) - \exp(-2t^2)| \leq 16\epsilon + |(\phi(t))^3\phi(-t) - \exp(-2t^2)|.$$

Since $\phi(t) = \exp(-\frac{1}{2}t^2) + h(t)$, therefore we get

$$|h(2t)| \leq 16\epsilon + \sum_{r=0}^{3}\binom{4}{r}\exp(-\frac{1}{2}rt^2)|h(t)|^{4-r}. \tag{6.20}$$

Put $t_n = \epsilon^{1/3}2^n$, where $n = 0, 1, 2, \ldots, [1 - \frac{2}{3}\log_2(\epsilon)]$, and let $h_n = \max\{|h(t)| : t_{n-1} \leq t \leq t^n\}$. Then (6.20) implies

$$h_{n+1} \leq 16\epsilon + 4\exp(-\frac{1}{2}t_n^2)h_n(1 + \frac{3}{2}h_n + h_n^2) + h_n^4. \tag{6.21}$$

Claim 6.3.1 *Relation (6.21) implies that for all sufficiently small $\epsilon > 0$ we have*

$$h_n \leq 2(C_0 + 44)\epsilon 4^n \exp(-t_0^2 4^n/6), \tag{6.22}$$

$$h_n^4 \leq \epsilon, \tag{6.23}$$

where $0 \leq n \leq [1 - \frac{2}{3}\log_2(\epsilon)]$, and C_0 is a constant from (6.19).

Claim 6.3.1 now ends the proof. Indeed,

$$\int_{-T}^{T} |\phi(t) - e^{-\frac{1}{2}t^2}|/t\,dt = 2\int_0^{t_0} |h(t)|/t\,dt + 2\sum_{i=1}^{n}\int_{t_{i-1}}^{t_i} |h(t)|/t\,dt$$

$$\leq 2C_0\epsilon + 2\sum_{i=1}^{n} h_i/t_{i-1}\int_{t_{i-1}}^{t_i} 1\,dt \leq 2C_0\epsilon + 4\sum_{i=1}^{n}(C_0 + 44)\epsilon 4^n e^{-t_0^2 4^n/6}$$

$$\leq 2C_0\epsilon + 24(C_0 + 44)\frac{\epsilon}{t_0^2}\int_0^{\infty} e^{-x}\,dx \leq C\epsilon^{1/3}.$$

$\square$

Proof of Claim 6.3.1. We shall prove (6.23) by induction, and (6.22) will be established in the induction step. By (6.19), inequality (6.23) is true for $n = 1$, provided $\epsilon < C_0^{-4/3}$. Suppose $m \geq 0$ is such that (6.23) holds for all $n \leq m$. Since $\frac{3}{2}h_n + h_n^2 < 3\epsilon 1/4 = \delta$, thus (6.21) implies

$$h_{m+1} \leq 32\epsilon + 4\exp(-\frac{1}{2}t_n^2)h_m(1 + \delta)$$

$$\leq 32\epsilon\sum_{j=1}^{n-1} 4^j(1 + \delta)^j \exp(-\frac{1}{2}\sum_{k=1}^{j} t_{n-k}^2) + 4^n(1 + \delta)^n \exp(-\frac{1}{2}\sum_{k=1}^{n} t_{n-k}^2)h_1$$

$$= 32\epsilon\sum_{j=1}^{n-1} 4^j(1 + \delta)^j \exp(-t_0^2(4^n - 4^{n-j})/6) + 4^n(1 + d)^n \exp(-t_0^2(4^n - 1)/6)h_1.$$

Therefore

$$h_{m+1} \leq (h_1 + 44\epsilon)(1 + \delta)^n 4^n e^{-t_0^2 4^n/6}. \tag{6.24}$$

Since

$$(1 + \delta)^n \leq (1 + 3\epsilon^{1/4})^{2 - \frac{2}{3}\log_2(\epsilon)} \leq 2$$

and

$$4^n e^{-t_0^2 4^n/6} \leq 4\epsilon^{-4/3}\exp(-\frac{1}{6}\epsilon^{-2/3}) \leq \epsilon^{-2/3}$$

for all $\epsilon > 0$ small enough, therefore, taking (6.19) into account, we get $h_{m+1} \leq 2(44 + C_0)\epsilon^{1/3} \leq \epsilon^{1/4}$, provided $\epsilon > 0$ is small enough. This proves (6.23) by induction. Inequality (6.22) follows now from (6.24). $\square$

6.4 Problems

Problem 6.1 *Show that for complex valued random variables X, Y*

$$|EXY - EXEY| \leq 16\alpha_{0,0}\|X\|_\infty\|Y\|_\infty.$$

(The constant is not sharp.)

Problem 6.2 *Suppose $(X, Y) \in \mathbb{R}^2$ are jointly normal and $\alpha_{0,1}(X, Y) < 1$. Show that X, Y are independent.*

Problem 6.3 *Suppose $(X, Y) \in \mathbb{R}^2$ are jointly normal with correlation coefficient ρ. Show that $Ef(X)g(Y) \leq \|f(X)\|_p\|g(Y)\|_p$ for all p-integrable $f(X), g(Y)$, provided $p \geq 1 + |\rho|$.*
Hint: *Use the explicit expression for conditional density and Hölder and Jensen inequalities.*

Problem 6.4 *Suppose $(X, Y) \in \mathbb{R}^2$ are jointly normal with correlation coefficient ρ. Show that*

$$corr(f(X), g(Y)) \leq |\rho|$$

for all square integrable $f(X), g(Y)$.

Problem 6.5 *Suppose $X, Y \in \mathbb{R}^2$ are jointly normal. Show that*

$$corr(f(X), g(Y)) \leq 2\pi\alpha_{0,0}(X, Y)$$

for all square integrable $f(X), g(Y)$.
Hint: *See Problem 2.3.*

Problem 6.6 *Let X, Y be random variables with finite moments of order $\alpha \geq 1$ and suppose that*

$$E\{|X - aY|^\alpha |Y\} \leq const;$$
$$E\{|Y - bX|^\alpha |X\} \leq const$$

for some real numbers a, b such that $ab \neq 0, 1, -1$. Show that X and Y have finite moments of all orders.

Problem 6.7 *Show that the conclusion of Theorem 6.2.2 can be strengthened to $E|X|^{|X|} < \infty$.*

Chapter 7

Conditional moments

In this chapter we shall use assumptions that mimic the behavior of conditional moments that would have followed from independence. Strictly speaking, corresponding characterization results do not generalize the theorems that assume independence, since weakening of independence is compensated by the assumption that the moments of appropriate order exist. However, besides just complementing the results of the previous chapters, the theory also has its own merits. Reference [37] points out the importance of description of probability distributions in terms of conditional distributions in statistical physics. From the mathematical point of view, the main advantage of conditional moments is that they are "less rigid" than the distribution assumptions. In particular, conditional moments lead to characterizations of some non-Gaussian distributions, see Problems 7.8 and 7.9.

The most natural conditional moment to use is, of course, the conditional expectation $E\{Z|\mathcal{F}\}$ itself. As in Section 4.1, we shall also use absolute conditional moments $E\{|Z|^\alpha|\mathcal{F}\}$, where α is a positive real number. Here we concentrate on $\alpha = 2$, which corresponds to the conditional variance. Recall that the conditional variance of a square-integrable random variable Z is defined by the formula

$$Var(Z|\mathcal{F}) = E\{(Z - E\{Z|\mathcal{F}\})^2|\mathcal{F}\} = E\{Z^2|\mathcal{F}\} - (E\{Z|\mathcal{F}\})^2.$$

7.1 Finite sequences

We begin with a simple result related to Theorem 5.1.1, compare [73, Theorem 5.3.2]; cf. also Problem 7.1 below.

Theorem 7.1.1 *If X_1, X_2 are independent identically distributed random variables with finite first moments, and for some $\alpha \neq 0, \pm 1$*

$$E\{X_1 - \alpha X_2|\alpha X_1 + X_2\} = 0, \tag{7.1}$$

then X_1 and X_2 are normal.

Proof. Let ϕ be a characteristic function of X_1. The joint characteristic function of the pair $X_1 - \alpha X_2, \alpha X_1 + X_2$ has the form $\phi(t + \alpha s)\phi(s - \alpha t)$. Hence, by Theorem 1.5.3,

$$\phi'(\alpha s)\phi(s) = \alpha\phi(\alpha s)\phi'(s).$$

Integrating this equation we obtain $\log \phi(\alpha s) = \alpha^2 \log \phi(s)$ in some neighborhood of 0.

If $\alpha^2 \neq 1$, this implies that $\phi(\alpha^{\pm n}) = \exp(C\alpha^{\pm 2n})$ for some complex constant C. This by Corollary 2.3.4 concludes the proof in each of the cases $0 < \alpha^2 < 1$ and $\alpha^2 > 1$ (in each of the cases one needs to choose the correct sign in the exponent of $\alpha^{\pm n}$). $\square$

Note that aside from the integrability condition, Theorem 7.1.1 resembles Theorem 5.1.1: clearly (7.1) follows if we assume that $X_1 - \alpha X_2$ and $\alpha X_1 + X_2$ are independent. There are however two major differences: parameter α is not allowed to take values ± 1, and X_1, X_2 are assumed to have equal distributions. We shall improve upon both in our Theorem 7.1.2 below. But we will use second order conditional moments, too.

The following result is a special but important case of a more difficult result [73, Theorem 5.7.1]; i. i. d. variant of the latter is given as Theorem 7.2.1 below.

Theorem 7.1.2 *Suppose X_1, X_2 are independent random variables with finite second moments such that*

$$E\{X_1 - X_2 | X_1 + X_2\} = 0, \tag{7.2}$$

$$E\{(X_1 - X_2)^2 | X_1 + X_2\} = const, \tag{7.3}$$

where const is a deterministic number. Then X_1 and X_2 are normal.

Proof. Without loss of generality, we may assume that X, Y are standardized random variables, ie. $EX = EY = 0, EX^2 = EY^2 = 1$ (the degenerate case is trivial). The joint characteristic function $\phi(t, s)$ of the pair $X+Y, X-Y$ equals $\phi_X(t+s)\phi_Y(t-s)$, where ϕ_X and ϕ_Y are the characteristic functions of X and Y respectively. Therefore by Theorem 1.5.3 condition (7.2) implies $\phi'_X(s)\phi_Y(s) = \phi_X(s)\phi'_Y(s)$. This in turn gives $\phi_Y(s) = \phi_X(s)$ for all real s close enough to 0.

Condition (7.3) by Theorem 1.5.3 after some arithmetics yields

$$\phi''_X(s)\phi_Y(s) + \phi_X(s)\phi''_Y(s) - 2\phi'_X(s)\phi'_Y(s) + 2\phi_X(s)\phi_Y(s) = 0.$$

This leads to the following differential equation for unknown function $\phi(s) = \phi_Y(s) = \phi_X(s)$

$$\phi''/\phi - (\phi'/\phi)^2 + 1 = 0, \tag{7.4}$$

valid in some neighborhood of 0. The solution of (7.4) with initial conditions $\phi''(0) = -1, \phi'(0) = 0$ is given by $\phi(s) = \exp(-\frac{1}{2}s^2)$, valid in some neighborhood of 0. By Corollary 2.3.4 this ends the proof of the theorem. $\square$

Remark: Theorem 7.1.2 also has Poisson, gamma, binomial and negative binomial distribution variants, see Problems 7.8 and 7.9.

Remark: The proof of Theorem 7.1.2 shows that for independent random variables condition (7.2) implies their characteristic functions are equal in a neighborhood of 0. Diaconis & Ylvisaker [36, Remark 1] give an example that the variables do not have to be equidistributed. (They also point out the relevance of this to statistics.)

7.2 Extension of Theorem 7.1.2

The next result is motivated by Theorem 5.3.1. Theorem 7.2.1 holds true also for non-identically distributed random variables, see eg. [73, Theorem 5.7.1] and is due to Laha [93, Corollary 4.1]; see also [101].

Theorem 7.2.1 *Let $X_1, \ldots, X_n$ be a sequence of square integrable independent identically distributed random variables and let $a_1, a_2, \ldots, a_n$, and $b_1, b_2, \ldots, b_n$ be given real numbers. Define random variables X, Y by $X = \sum_{k=1}^n a_k X_k, Y = \sum_{k=1}^n b_k X_k$ and suppose that for some constants ρ, α we have*

$$E\{X|Y\} = \rho Y + \alpha \tag{7.5}$$

and

$$Var(X|Y) = const, \tag{7.6}$$

where const is a deterministic number. If for some $1 \leq k \leq n$ we have $a_k - \rho b_k \neq 0$, then X_1 is Gaussian.

Lemma 7.2.2 *Under the assumptions of Theorem 7.2.1, all moments of random variable X_1 are finite.*

Indeed, consider $N(x) = P(|X_1| \geq x)$. Clearly without loss of generality we may assume $a_1 b_1 \neq 0$. Event $\{|X_1| \geq Cx\}$, where C is a (large) constant to be chosen later, can be decomposed into the sum of disjoint events

$$A = \{|X_1| \geq Cx\} \cap \bigcup_{j=2}^n \{|X_j| \geq x\}$$

and

$$B = \{|X_1| \geq Cx\} \cap \bigcap_{j=2}^n \{|X_j| < x\}.$$

Since $P(A) \leq (n-1)P(|X_1| \geq Cx, |X_2| \geq x)$, therefore $P(|X_1| \geq Cx) \leq P(A) + P(B) \leq nN^2(x) + P(B)$.

Clearly, if $|X_1| \geq Cx$ and all other $|X_j| < x$, then $|\sum a_k X_k - \rho \sum b_k X_k| \geq (C|a_1 - \rho b_1| - \sum |a_k - \rho b_k|)x$ and similarly $|\sum b_k X_k| \geq (C|b_1| - \sum |b_j|)x$. Hence we can find constants $C_1, C_2 > 0$ such that

$$P(B) \leq P(|X - \rho Y| > C_1 x, |Y| > C_2 x).$$

Using conditional version of Chebyshev's inequality and (7.6) we get

$$N(Cx) \leq nN^2(x) + C_3 N(C_2 x)/x^2. \tag{7.7}$$

This implies that moments of all orders are finite, see Corollary 1.3.4. Indeed, since $N(x) \leq C/x^2$, inequality (7.7) implies that there are $K < \infty$ and $\epsilon > 0$ such that

$$N(x) \leq KN(\epsilon x)/x^2$$

for all large enough x.

Proof of Theorem 7.2.1. Without loss of generality, we shall assume that $EX_1 = 0$ and $Var(X_1) = 1$. Then $\alpha = 0$. Let $Q(t)$ be the logarithm of the characteristic function of X_1, defined in some neighborhood of 0. Equation (7.5) and Theorem 1.5.3 imply

$$\sum a_k Q'(tb_k) = \rho \sum b_k Q'(tb_k). \tag{7.8}$$

Similarly (7.6) implies

$$\sum a_k^2 Q''(tb_k) = -\sigma^2 + \rho^2 \sum b_k^2 Q''(tb_k). \tag{7.9}$$

Differentiating (7.8) we get

$$\sum a_k b_k Q''(tb_k) = \rho \sum b_k^2 Q''(tb_k),$$

which multiplied by 2ρ and subtracted from (7.9) gives after some calculation

$$\sum (a_k - \rho b_k)^2 Q''(tb_k) = -\sigma^2. \tag{7.10}$$

Lemma 7.2.2 shows that all moments of X exist. Therefore, differentiating (7.10) we obtain

$$\sum (a_k - \rho b_k)^2 b_k^{2r} Q^{(2r+2)}(0) = 0 \tag{7.11}$$

for all $r \geq 1$.

This shows that $Q^{(2r+2)}(0) = 0$ for all $r \geq 1$. The characteristic function ϕ of random variable $X_1 - X_2$ satisfies $\phi(t) = \exp(2 \sum_r t^{2r} Q^{(2r)}(0)/(2r)!)$; hence by Theorem 2.5.1 it corresponds to the normal distribution. By Theorem 2.5.2, X_1 is normal. $\square$

Remark: Lemma 7.2.2 can be easily extended to non-identically distributed random variables.

7.3 Application: the Central Limit Theorem

In this section we shall show how the characterization of the normal distribution might be used to prove the Central Limit Theorem. The following is closely related to [10, Theorem 19.4].

Theorem 7.3.1 *Suppose that pairs (X_n, Y_n) converge in distribution to independent r. v. (X, Y). Assume that*

(a) $\{X_n^2\}$ and $\{Y_n^2\}$ are uniformly integrable;

(b) $E\{X_n | X_n + Y_n\} - 2^{-1/2}(X_n + Y_n) \to 0$ in L_1 as $n \to \infty$;

(c)

$$Var(X_n | X_n + Y_n) \to 1/2 \text{ in } L_1 \text{ as } n \to \infty. \tag{7.12}$$

Then X is normal.

Our starting point is the following variant of Theorem 7.2.1.

Lemma 7.3.2 *Suppose X, Y are nondegenerate (ie. $EX^2 EY^2 \neq 0$) centered independent random variables. If there are constants c, K such that*

$$E\{X|X + Y\} = c(X + Y) \tag{7.13}$$

and

$$Var(X|X + Y) = K, \tag{7.14}$$

then X and Y are normal.

Proof. Let Q_X, Q_Y denote the logarithms of the characteristic functions of X, Y respectively. By Theorem 1.5.3 (see also Problem 1.19, with $Q(t, s) = Q_X(t + s) + Q_Y(s)$), equation (7.13) implies

$$(1 - c)Q'_X(s) = cQ'_Y(s) \tag{7.15}$$

for all s close enough to 0.

Differentiating (7.15) we see that $c = 0$ implies $EX^2 = 0$; similarly, $c = 1$ implies $Y = 0$. Therefore, without loss of generality we may assume $c(1 - c) \neq 0$ and $Q_X(s) = C_1 + C_2 Q_Y(s)$ with $C_2 = c/(1 - c)$.

From (7.14) we get

$$Q''_X(s) = -K + c^2(Q''_X(s) + Q''_Y(s)),$$

which together with (7.15) implies $Q''_Y(s) = const.$ $\square$

Proof of Theorem 7.3.1. By uniform integrability, the limiting r. v. X, Y satisfy the assumption of Lemma 7.3.2. This can be easily seen from Theorem 1.5.3 and (1.18), see also Problem 1.21. Therefore the conclusion follows. $\square$

7.3.1 CLT for i. i. d. sums

Here is the simplest application of Theorem 7.3.1.

Theorem 7.3.3 *Suppose ξ_j are centered i. i. d. with $E\xi^2 = 1$. Put $S_n = \sum_{j=1}^{n} \xi_j$. Then $\frac{1}{\sqrt{n}} S_n$ is asymptotically $N(0,1)$ as $n \to \infty$.*

Proof. We shall show that every convergent in distribution subsequence converges to $N(0, 1)$. Having bounded variances, pairs $(\frac{1}{\sqrt{n}} S_n, \frac{1}{\sqrt{n}} S_{2n})$ are tight and one can select a subsequence n_k such that both components converge (jointly) in distribution. We shall apply Theorem 7.3.1 to $X_k = \frac{1}{\sqrt{n_k}} S_{n_k}, X_k + Y_k = \frac{1}{\sqrt{n_k}} S_{2n_k}$.

(a) The i. i. d. assumption implies that $\frac{1}{n} S_n^2$ are uniformly integrable, cf. Proposition 1.7.1. The fact that the limiting variables (X, Y) are independent is obvious as X, Y arise from sums over *disjoint blocks*.

(b) $E\{S_n|S_{2n}\} = \frac{1}{2} S_{2n}$ by symmetry, see Problem 1.11.

(c) To verify (7.12) notice that $S_n^2 = \sum_{j=1}^{n} \xi_j^2 + \sum_{k \neq j} \xi_j \xi_k$. By symmetry

$$E\{\xi_1^2 | S_{2n}, \sum_{j=1}^{2n} \xi_j^2\}$$

$$= \frac{1}{2n} \sum_{j=1}^{2n} \xi_j^2$$

and

$$E\{\xi_1 \xi_2 | S_{2n}, \sum_{k \neq j,\ k,j \leq 2n} \xi_j \xi_k\}$$

$$= \frac{1}{2n(2n-1)} \sum_{k \neq j,\ k,j \leq 2n} \xi_j \xi_k = \frac{1}{2n(2n-1)}(S_{2n}^2 - \sum_{j=1}^{2n} \xi_j^2).$$

Therefore

$$Var(S_n | S_{2n}) = \frac{n}{4n-2} E\{\sum_{j=1}^{2n} \xi_j^2 | S_{2n}\} - \frac{1}{4(2n-1)} S_{2n}^2.$$

Since $\frac{1}{n} \sum_{j=1}^{n} \xi_j^2 \to 1$ in L_1, this implies (7.12). $\square$

7.4 Application: independence of empirical mean and variance

For a normal distribution it is well known that the empirical mean and the empirical variance are independent. The next result gives a converse implication; our proof is a version of the proof sketched in [73, Remark on page 103], who give also a version for non-identically distributed random variables.

Theorem 7.4.1 *Let $X_1, \ldots, X_n$ be i. i. d. and denote $\bar{X} = \frac{1}{n} \sum_{j=1}^{n} X_j$, $S^2 = \frac{1}{n} \sum_{j=1}^{n} X_j^2 - \bar{X}^2$. If $n \geq 2$ and $\bar{X}, S^2$ are independent, then X_1 is normal.*

The following lemma resembles Lemma 5.3.2 and replaces [73, Theorem 4.3.1].

Lemma 7.4.2 *Under the assumption of Theorem 7.4.1, the moments of X_1 are finite.*

Proof. Let $q = (2n)^{-1}$. Then

$$P(|X_1| > t) \tag{7.16}$$
$$\leq \sum_{j=2}^{n} P(|X_1| > t, |X_j| > qt) + P(|X_1| > t, |X_2| \leq qt, \ldots, |X_n| \leq qt).$$

Clearly, one can find T such that

$$\sum_{j=2}^{n} P(|X_1| > t, |X_j| > qt)$$

$$= (n-1)P(|X_1| > t)P(|X_1| > qt) \leq \frac{1}{2} P(|X_1| > t)$$

for all $t > T$. Therefore

$$P(|X_1| > t) \leq 2P(|X_1| > t, |X_2| \leq qt, \ldots, |X_n| \leq qt). \tag{7.17}$$

Event $\{|X_1| > t, |X_2| \le qt, \ldots, |X_n| \le qt\}$ implies $|\bar{X}| > (1 - nq)t/n$. It also implies $S^2 > \frac{1}{n}(X_1 - \bar{X})^2 > \frac{1}{4n}t^2$. Therefore by independence

$$P(|X_1| > t, |X_2| \le qt, \ldots, |X_n| \le qt)$$

$$\le P\left(|\bar{X}| > \frac{1}{2n}t\right) P\left(S^2 > \frac{1}{4n}t^2\right)$$

$$\le nP\left(|X_1| > \frac{1}{2n}t\right) P\left(S^2 > \frac{1}{4n}t^2\right).$$

$$\le P\left(|\bar{X}| > \frac{1}{2n}t\right) P\left(S^2 > \frac{1}{4n}t^2\right)$$

$$\le nP\left(|X_1| > \frac{1}{2n}t\right) P\left(S^2 > \frac{1}{4n}t^2\right).$$

This by (7.17) and Corollary 1.3.3 ends the proof. Indeed, $n \ge 2$ is fixed and $P(S^2 > \frac{1}{4n}t^2)$ is arbitrarily small for large t. $\square$

Proof of Theorem 7.4.1. By Lemma 7.4.2, the second moments are finite. Therefore the independence assumption implies that the corresponding conditional moments are constant. We shall apply Lemma 7.3.2 with $X = X_1$ and $Y = \sum_{j=2}^{n} X_j$.

The assumptions of this lemma can be quickly verified as follows. Clearly, $E\{X_1|\bar{X}\} = \bar{X}$ by i. i. d., proving (7.13). To verify (7.14), notice that again by symmetry (i. i. d.)

$$E\{X_1^2|\bar{X}\} = E\{\frac{1}{n}\sum_{j=1}^{n} X_j^2|\bar{X}\} = E\{S^2|\bar{X}\} + \bar{X}^2.$$

By independence, $E\{S^2|\bar{X}\} = ES^2 = const$, verifying (7.14) with $K = const$. $\square$

7.5 Infinite sequences and conditional moments

In this section we present results that hold true for infinite sequences only; they fail for finite sequences. We consider assumptions that involve first two conditional moments only. They resemble (7.5) and (7.6) but, surprisingly, independence assumption can be omitted when infinite sequences are considered.

To simplify the notation, we limit our attention to L_2-homogeneous Markov chains only. A similar non-Markovian result will be given in Section 8.3 below. Problem 7.7 shows that Theorem 7.5.1 is not valid for finite sequences.

Theorem 7.5.1 *Let $X_1, X_2, \ldots$ be an infinite Markov chain with finite and non-zero variances and assume that there are numbers $c_1 = c_1(n), \ldots, c_7 = c_7(n)$, such that the following conditions hold for all $n = 1, 2, \ldots$*

$$E\{X_{n+1}|X_n\} = c_1 X_n + c_2, \tag{7.18}$$

$$E\{X_{n+1}|X_n, X_{n+2}\} = c_3 X_n + c_4 X_{n+2} + c_5, \tag{7.19}$$

$$Var(X_{n+1}|X_n) = c_6, \qquad\qquad (7.20)$$

$$Var(X_{n+1}|X_n, X_{n+2}) = c_7. \qquad\qquad (7.21)$$

Furthermore, suppose that correlation coefficient $\rho = \rho(n)$ between random variables X_n and X_{n+1} does not depend on n and $\rho^2 \neq 0, 1$. Then (X_k) is a Gaussian sequence.

Notation for the proof. Without loss of generality we may assume that each X_n is a standardized random variable, ie. $EX_n = 0, EX_n^2 = 1$. Then it is easily seen that $c_1 = \rho$, $c_2 = c_5 = 0$, $c_3 = c_4 = \rho/(1 + \rho^2)$, $c_6 = 1 - \rho^2$, $c_7 = (1 - \rho^2)/(1 + \rho^2)$. For instance, let us show how to obtain the expression for the first two constants c_1, c_2. Taking the expected value of (7.18) we get $c_2 = 0$. Then multiplying (7.18) by X_n and taking the expected value again we get $EX_n X_{n+1} = c_1 EX_n^2$. Calculation of the remaining coefficients is based on similar manipulations and the formula $EX_n X_{n+k} = \rho^k$; the latter follows from (7.18) and the Markov property. For instance,

$$c_7 = EX_{n+1}^2 - (\rho/(1 + \rho^2))^2 E(X_n + X_{n+2})^2 = 1 - 2\rho^2/(1 + \rho^2).$$

The first step in the proof is to show that moments of all orders of $X_n, n = 1, 2, \ldots$ are finite. If one is willing to add the assumptions reversing the roles of n and $n + 1$ in (7.18) and (7.20), then this follows immediately from Theorem 6.2.2 and there is no need to restrict our attention to L_2-homogeneous chains. In general, some additional work needs to be done.

Lemma 7.5.2 *Moments of all orders of $X_n, n = 1, 2, \ldots$ are finite.*

Sketch of the proof: Put $X = X_n, Y = X_{n+1}$, where $n \geq 1$ is fixed. We shall use Theorem 6.2.2. From (7.18) and (7.20) it follows that (6.7) is satisfied. To see that (6.6) holds, it suffices to show that $E\{X|Y\} = \rho Y$ and $Var(X|Y) = 1 - \rho^2$.

To this end, we show by induction that

$$E\{X_{n+r}|X_n, X_{n+k}\} = a_{k,r} X_n + b_{k,r} X_{n+k} \qquad\qquad (7.22)$$

is linear for $0 \leq r \leq k$.

Once (7.22) is established, constants can easily be computed analogously to computation of c_j in (7.18) and (7.19). Multiplying the last equality by X_n, then by X_{n+k} and taking the expectations, we get $b_{k,r} = \frac{\rho^{k-r} - \rho^{k+r}}{1 - \rho^{2k}}$ and $a_{k,r} = \rho^r - b_{k,r}\rho^k$.

The induction proof of (7.22) goes as follows. By (7.19) the formula is true for $k = 2$ and all $n \geq 1$. Suppose (7.22) holds for some $k \geq 2$ and all $n \geq 1$. By the Markov property

$$E\{X_{n+r}|X_n, X_{n+k+1}\}$$

$$= E^{X_n, X_{n+k+1}} E\{X_{n+r}|X_n, X_{n+k}\}$$

$$= a_{k,r} X_n + b_{k,r} E\{X_{n+k}|X_n, X_{n+k+1}\}.$$

This reduces the proof to establishing the linearity of $E\{X_{n+k}|X_n, X_{n+k+1}\}$.

We now concentrate on the latter. By the Markov property, we have

$$E\{X_{n+k}|X_n, X_{n+k+1}\}$$

$$= E^{X_n, X_{n+k+1}} E\{X_{n+k}|X_{n+1}, X_{n+k+1}\}$$

$$= b_{k+1,1} X_{n+k+1} + a_{k+1,1} E\{X_{n+1}|X_n, X_{n+k+1}\}.$$

We have $a_{k+1,1} = \rho^{k-1}\frac{1-\rho^2}{1-\rho^{2k}}$; in particular, $a_{k+1,1} = \sinh(\log\rho)/\sinh(k\log\rho)$ so that one can easily see that $0 < a_{k+1,1} < 1$. Since

$$E\{X_{n+1}|X_n, X_{n+k+1}\} = E^{X_n, X_{n+k+1}} E\{X_{n+1}|X_n, X_{n+k}\}$$

$$= a_{k,1} X_n + b_{k,1} E\{X_{n+k}|X_n, X_{n+k+1}\},$$

we get

$$E\{X_{n+k}|X_n, X_{n+k+1}\} \tag{7.23}$$
$$= b_{k+1,1} X_{n+k+1} + a_{k+1,1} a_{k,1} X_n + a_{k+1,1} b_{k,1} E\{X_{n+k}|X_n, X_{n+k+1}\}.$$

Notice that $b_{k,1} = \rho^{k-1}\frac{1-\rho^2}{1-\rho^{2k}} = a_{k+1,1}$. In particular $0 < a_{k+1,1}b_{k,1} < 1$. Therefore (7.23) determines $E\{X_{n+k}|X_n, X_{n+k+1}\}$ uniquely as a linear function of X_n and X_{n+k+1}. This ends the proof of (7.22).

Explicit formulas for coefficients in (7.22) show that $b_{k,1} \to 0$ and $a_{k,1} \to \rho$ as $k \to \infty$. Applying conditional expectation $E^{X_n}\{.\}$ to (7.22) we get $E\{X_{n+1}|X_n\} = \lim_{k\to\infty}(aX_n + bE\{X_{n+k}|X_n\}) = \rho X_n$, which establishes required $E\{X|Y\} = \rho Y$.

Similarly, we check by induction that

$$Var(X_{n+r}|X_n, X_{n+k}) = c \tag{7.24}$$

is non-random for $0 \le r \le k$; here c is computed by taking the expectation of (7.24); as in the previous case, c depends on ρ, r, k.

Indeed, by (7.21) formula (7.24) holds true for $k = 2$. Suppose it is true for some $k \ge 2$, ie. $E\{X_{n+r}^2|X_n, X_{n+k}\} = c + (a_k X_n + b_k X_{n+k})^2$, where $a_k = a_{k,r}$, $b_k = b_{k,r}$ come from (7.22). Then

$$E\{X_{n+r}^2|X_n, X_{n+k+1}\}$$

$$= E^{X_n, X_{n+k+1}} E\{X_{n+k}^2|X_{n+1}, X_{n+k}\}$$

$$= c + E^{X_n, X_{n+k+1}}(a_k X_n + b_k X_{n+k})^2$$

$$= b^2 E^{X_n, X_{n+k+1}}\{X_{n+k}^2\} + \textit{quadratic polynomial in } X_n.$$

We write again

$$E\{X_{n+k}^2|X_n, X_{n+k+1}\}$$

$$= E^{X_n, X_{n+k+1}} E\{X_{n+k}^2|X_{n+1}, X_{n+k+1}\}$$

$$= b^2 E\{X_{n+1}^2|X_n, X_{n+k+1}\} + \textit{quadratic polynomial in } X_{n+k+1}.$$

Since

$$E\{X_{n+1}^2|X_n, X_{n+k+1}\}$$

$$= E^{X_n, X_{n+k+1}} E\{X_{n+1}^2|X_n, X_{n+k}\}$$

$$= a^2 E\{X_{n+k}^2|X_n, X_{n+k+1}\} + \textit{quadratic polynomial in } X_n$$

and since $a^2 b^2 \ne 1$ (those are the same coefficients that were used in the first part of the proof; namely, $a = a_{k+1,1}$, $b = b_{k,1}$.) we establish that $E\{X_{n+r}^2|X_n, X_{n+k}\}$ is a quadratic polynomial in variables X_n, X_{n+k}. A more careful analysis permits to recover the coefficients of this polynomial to see that actually (7.24) holds.

This shows that (6.6) holds and by Theorem 6.2.2 all the moments of $X_n, n \geq 1$, are finite. $\Box$

We shall prove Theorem 7.5.1 by showing that all mixed moments of $\{X_n\}$ are equal to the corresponding moments of a suitable Gaussian sequence. To this end let $\vec{\gamma} = (\gamma_1, \gamma_2, \ldots)$ be the mean zero Gaussian sequence with covariances $E\gamma_i\gamma_j$ equal to EX_iX_j for all $i, j \geq 1$. It is well known that the sequence $\gamma_1, \gamma_2, \ldots$ satisfies (7.18)–(7.21) with the same constants $c_1, \ldots, c_7$, see Theorem 2.2.9. Moreover, $(\gamma_1, \gamma_2, \ldots)$ is a Markov chain, too.

We shall use the variant of the method of moments.

Lemma 7.5.3 *If $\mathbf{X} = (X_1, \ldots, X_d)$ is a random vector such that all moments*

$$EX_1^{i(1)} \ldots X_d^{i(d)} = E\gamma_1^{i(1)} \ldots \gamma_d^{i(d)}$$

are finite and equal to the corresponding moments of a multivariate normal random variable $\mathbf{Z} = (\gamma_1, \ldots, \gamma_d)$, then $\mathbf{X}$ and $\mathbf{Z}$ have the same (normal) distribution.

Proof. It suffice to show that $E\exp(i\mathbf{t} \cdot \mathbf{X}) = E\exp(i\mathbf{t} \cdot \mathbf{Z})$ for all $\mathbf{t} \in \mathbb{R}^d$ and all $d \geq 1$. Clearly, the moments of $(\mathbf{t} \cdot \mathbf{X})$ are given by

$$E(\mathbf{t} \cdot \mathbf{X})^k = \sum_{i(1)+\ldots+i(d)=k} t_1^{i(1)} \ldots t_d^{i(d)} EX_1^{i(1)} \ldots X_d^{i(d)}$$

$$= \sum_{i(1)+\ldots+i(d)=k} t_1^{i(1)} \ldots t_d^{i(d)} E\gamma_1^{i(1)} \ldots \gamma_d^{i(d)}$$

$$= E(\mathbf{t} \cdot \mathbf{Z})^k, k = 1, 2, \ldots$$

One dimensional random variable $(\mathbf{t} \cdot \mathbf{X})$ satisfies the assumption of Corollary 2.3.3; thus $E\exp(it \cdot \mathbf{X}) = E\exp(it \cdot \mathbf{Z})$, which ends the proof. $\Box$

The main difficulty in the proof is to show that the appropriate higher centered conditional moments are the same for both sequences $\mathbf{X}$ and $\vec{\gamma}$; this is established in Lemma 7.5.4 below. Once Lemma 7.5.4 is proved, all mixed moments can be calculated easily (see Lemma 7.5.5 below) and Lemma 7.5.3 will end the proof.

Lemma 7.5.4 *Put $X_0 = \gamma_0 = 0$. Then*

$$E\{(X_n - \rho X_{n-1})^k | X_{n-1}\} = E\{(\gamma_n - \rho\gamma_{n-1})^k | \gamma_{n-1}\} \tag{7.25}$$

for all $n, k = 1, 2 \ldots$

Proof. We shall show simultaneously that (7.25) holds and that

$$E\{(X_{n+1} - \rho^2 X_{n-1})^k | X_{n-1}\} = E\{(\gamma_{n+1} - \rho^2\gamma_{n-1})^k | \gamma_{n-1}\} \tag{7.26}$$

for all $n, k = 1, 2 \ldots$. The proof of (7.25) and (7.26) is by induction with respect to parameter k. By our choice of $(\gamma_1, \gamma_2, \ldots)$, formula (7.25) holds for all n and for the first two conditional moments, ie. for $k = 0, 1, 2$. Formula (7.26) is also easily seen to hold for $k = 1$; indeed, from the Markov property $E\{X_{n+1} - \rho^2 X_{n-1}|X_{n-1}\} = E\{E\{X_{n+1}|X_n\}|X_{n-1}\} - \rho^2 X_{n-1} = 0$. We now check that (7.26) holds for $k = 2$, too. This goes by simple re-arrangement, the Markov property and (7.20):

$$E\{(X_{n+1} - \rho^2 X_{n-1})^2|X_{n-1}\}$$

$$= E\{(X_{n+1} - \rho X_n)^2 + \rho^2(X_n - \rho X_{n-1})^2 + 2\rho(X_n - \rho X_{n-1})(X_{n+1} - \rho X_n)|X_{n-1}\}$$

$$= E\{(X_{n+1} - \rho X_n)^2|X_{n-1}\} + \rho^2 E\{(X_n - \rho X_{n-1})^2|X_{n-1}\}$$

$$= E\{(X_{n+1} - \rho X_n)^2|X_n\} + \rho^2 E\{(X_n - \rho X_{n-1})^2|X_{n-1}\} = 1 - \rho^4.$$

Since the same computation can be carried out for the Gaussian sequence (γ_k), this establishes (7.26) for $k = 2$.

Now we continue the induction part of the proof. Suppose (7.25) and (7.26) hold for all n and all $k \leq m$, where $m \geq 2$. We are going to show that (7.25) and (7.26) hold for $k = m + 1$ and all $n \geq 1$. This will be established by keeping $n \geq 1$ fixed and producing a system of two linear equations for the two unknown conditional moments

$$x = E\{(X_{n+1} - \rho^2 X_{n-1})^{m+1}|X_{n-1}\}$$

and

$$y = E\{(X_n - \rho X_{n-1})^{m+1}|X_{n-1}\}.$$

Clearly, x, y are random; all the identities below hold with probability one.

To obtain the first equation, consider the expression

$$W = E\{(X_n - \rho X_{n-1})(X_{n+1} - \rho^2 X_{n-1})^m|X_{n-1}\}. \tag{7.27}$$

We have

$$E\{(X_n - \rho X_{n-1})(X_{n+1} - \rho X_n)^m|X_{n-1}\}$$

$$= E\{E\{X_n - \rho X_{n-1}|X_{n-1}, X_{n+1}\}(X_{n+1} - \rho^2 X_{n-1})^m|X_{n-1}\}.$$

Since by (7.19)

$$E\{X_n - \rho X_{n-1}|X_{n-1}, X_{n+1}\} = \rho/(1 + \rho^2)(X_{n+1} - \rho^2 X_{n-1}),$$

hence

$$W = \rho/(1 + \rho^2)E\{(X_{n+1} - \rho^2 X_{n-1})^{m+1}|X_{n-1}\}. \tag{7.28}$$

On the other hand we can write

$$W = E\left\{(X_n - \rho X_{n-1})\left((X_{n+1} - \rho X_n) + \rho(X_n - \rho X_{n-1})\right)^m|X_{n-1}\right\}.$$

By the binomial expansion

$$\left((X_{n+1} - \rho X_n) + \rho(X_n - \rho X_{n-1})\right)^m$$

$$= \sum_{k=0}^m \binom{m}{k} \rho^k (X_{n+1} - \rho X_n)^{m-k}(X_n - \rho X_{n-1})^k.$$

Therefore the Markov property gives

$$W = \sum_{k=0}^{m} \binom{m}{k} \rho^k E\left\{ (X_n - \rho X_{n-1})^{k+1} E\{(X_{n+1} - \rho X_n)^{m-k}|X_n\} |X_{n-1}\right\}$$

$$= \rho^m E\{(X_n - \rho X_{n-1})^{m+1}|X_{n-1}\} + R,$$

where

$$R = \sum_{k=0}^{m-1} \binom{m}{k} \rho^k E\left\{ (X_n - \rho X_{n-1})^{k+1} E\{(X_{n+1} - \rho X_n)^{m-k}|X_n\} |X_{n-1}\right\}$$

is a deterministic number, since $E\{(X_{n+1} - \rho X_n)^{m-k}|X_n\}$ and $E\{(X_n - \rho X_{n-1})^{k+1}|X_{n-1}\}$ are uniquely determined and non-random for $0 \le k \le m - 1$.

Comparing this with (7.28) we get the first equation

$$\rho/(1 + \rho^2)x = \rho^m y + R \tag{7.29}$$

for the unknown (and at this moment yet random) x and y.

To obtain the second equation, consider

$$V = E\{(X_n - \rho X_{n-1})^2 (X_{n+1} - \rho^2 X_{n-1})^{m-1}|X_{n-1}\}. \tag{7.30}$$

We have

$$E\{(X_n - \rho X_{n-1})^2 (X_{n+1} - \rho^2 X_{n-1})^{m-1}|X_{n-1}\}$$

$$= E\left\{ E\{(X_n - \rho X_{n-1})^2|X_{n-1}, X_{n+1}\}(X_{n+1} - \rho^2 X_{n-1})^{m-1} |X_{n-1}\right\}.$$

Since

$$X_n - \rho X_{n-1}$$

$$= X_n - \rho/(1 + \rho^2)(X_{n+1} + X_{n-1}) + \rho/(1 + \rho^2)(X_{n+1} - \rho^2 X_{n-1}),$$

by (7.19) and (7.21) we get

$$E\{(X_n - \rho X_{n-1})^2|X_{n-1}, X_{n+1}\}$$

$$= (1 - \rho^2)/(1 + \rho^2) + \left(\rho/(1 + \rho^2)(X_{n+1} - \rho^2 X_{n-1})\right)^2.$$

Hence

$$V = \left(\rho/(1 + \rho^2)\right)^2 E\{(X_{n+1} - \rho^2 X_{n-1})^{m+1}|X_{n-1}\} + R', \tag{7.31}$$

where by induction assumption $R' = c_7 E\{(X_{n+1} - \rho^2 X_{n-1})^{m-1}|X_{n-1}\}$ is uniquely determined and non-random. On the other hand we have

$$V = E\left\{ (X_n - \rho X_{n-1})^2 \left((X_{n+1} - \rho X_n) + \rho(X_n - \rho X_{n-1})\right)^{m-1} |X_{n-1}\right\}.$$

By the binomial expansion

$$\left((X_{n+1} - \rho X_n) + \rho(X_n - \rho X_{n-1})\right)^{m-1}$$

$$= \sum_{k=0}^{m-1} \binom{m-1}{k} (X_{n+1} - \rho X_n)^{m-k-1}(X_n - \rho X_{n-1})^k.$$

Therefore the Markov property gives

$$V =$$

$$\sum_{k=0}^{m-1} \binom{m-1}{k} \rho^k E\left\{(X_n - \rho X_{n-1})^{k+2} E\{(X_{n+1} - \rho X_n)^{m-k-1}|X_n\} \Big| X_{n-1}\right\}$$

$$= \rho^{m-1} E\{(X_n - \rho X_{n-1})^{m+1}|X_{n-1}\} + R'',$$

where

$$R'' =$$

$$\sum_{k=0}^{m-2} \binom{m-1}{k} \rho^k E\left\{(X_n - \rho X_{n-1})^{k+2} E\{(X_{n+1} - \rho X_n)^{m-k-1}|X_n\} \Big| X_{n-1}\right\}$$

is a non-random number, since by induction assumption, for $0 \leq k \leq m - 2$ both $E\{(X_{n+1} - \rho X_n)^{m-k-1}|X_n\}$ and $E\{(X_n - \rho X_{n-1})^{k+2}|X_{n-1}\}$ are uniquely determined non-random numbers.

Equating both expression for V gives the second equation:

$$\rho^{m-1}y = (\frac{\rho}{1 + \rho^2})^2 x + R_1, \tag{7.32}$$

where again R_1 is uniquely determined and non-random.

The determinant of the system of two linear equations (7.29) and (7.32) is $\rho^m/(1 + \rho^2)^2 \neq 0$. Therefore conditional moments x, y are determined uniquely. In particular, they are equal to the corresponding moments of the Gaussian distribution and are non-random. This ends the induction, and the lemma is proved. $\square$

Lemma 7.5.5 *Equalities (7.25) imply that* $\mathbf{X}$ *and* $\vec{\gamma}$ *have the same distribution.*

Proof. By Lemma 7.5.3, it remains to show that

$$E X_1^{i(1)} \ldots X_d^{i(d)} = E \gamma_1^{i(1)} \ldots \gamma_d^{i(d)} \tag{7.33}$$

for every $d \geq 1$ and all $i(1), \ldots, i(d) \in \mathbb{N}$. We shall prove (7.33) by induction with respect to d. Since $EX_i = 0$ and $E\{\cdot|X_0\} = E\{\cdot\}$, therefore (7.33) for $d = 1$ follows immediately from (7.25).

If (7.33) holds for some $d \geq 1$, then write $X_{d+1} = (X_{d+1} - \rho X_d) + \rho X_d$. By the binomial expansion

$$E X_1^{i(1)} \ldots X_d^{i(d)} X_{d+1}^{i(d+1)} \tag{7.34}$$

$$= \sum_{j=0}^{i(d+1)} \binom{i(d+1)}{j} \rho^{i(d+1)-j} E X_1^{i(1)} \ldots X_d^{i(d)} E\{(X_{d+1} - \rho X_d)^j|X_d\} X_d^{i(d+1)-j}.$$

Since by assumption

$$E\{(X_{d+1} - \rho X_d)^j|X_d\} = E\{(\gamma_{d+1} - \rho \gamma_d)^j|\gamma_d\}$$

is a deterministic number for each $j \geq 0$, and since by induction assumption

$$E X_1^{i(1)} \ldots X_d^{i(d)+i(d+1)-j} = E \gamma_1^{i(1)} \ldots \gamma_d^{i(d)+i(d+1)-j},$$

therefore (7.34) ends the proof. $\square$

7.6   Problems

Problem 7.1 *Show that if X_1 and X_2 are i. i. d. then*

$$E\{X_1 - X_2 | X_1 + X_2\} = 0.$$

Problem 7.2 *Show that if X_1 and X_2 are i. i. d. symmetric and*

$$E\{(X_1 + X_2)^2 | X_1 - X_2\} = const,$$

then X_1 is normal.

Problem 7.3 *Show that if X, Y are independent integrable and $E\{X|X+Y\} = EX$ then $X = const$.*

Problem 7.4 *Show that if X, Y are independent integrable and $E\{X|X+Y\} = X+Y$ then $Y = 0$.*

Problem 7.5 ([36]) *Suppose X, Y are independent, X is nondegenerate, Y is integrable, and $E\{Y|X+Y\} = a(X+Y)$ for some a.*
(i) Show that $|a| \leq 1$.
(ii) Show that if $E|X|^p < \infty$ for some $p > 1$, then $E|Y|^p < \infty$. Hint By Problem 7.3, $a \neq 1$.

Problem 7.6 ([36, page 122]) *Suppose X, Y are independent, X is nondegenerate normal, Y is integrable, and $E\{Y|X+Y\} = a(X+Y)$ for some a.*
Show that Y is normal.

Problem 7.7 *Let X, Y be (dependent) symmetric random variables taking values ± 1. Fix $0 \leq \theta \leq 1/2$ and choose their joint distribution as follows.*

$$P_{X,Y}(-1, 1) = 1/2 - \theta,$$

$$P_{X,Y}(1, -1) = 1/2 - \theta,$$

$$P_{X,Y}(-1, -1) = 1/2 + \theta,$$

$$P_{X,Y}(1, 1) = 1/2 + \theta.$$

Show that

$$E\{X|Y\} = \rho Y \ \text{ and } \ E\{Y|X\} = \rho Y;$$

$$Var(X|Y) = 1 - \rho^2 \ \text{ and } \ Var(Y|X) = 1 - \rho^2$$

and the correlation coefficient satisfies $\rho \neq 0, \pm 1$.

Problems below characterize some non-Gaussian distributions, see [12, 131, 148].

Problem 7.8 *Prove the following variant of Theorem 7.1.2:*

If X_1, X_2 are i. i. d. random variables with finite second moments, and

$$Var(X_1 - X_2 | X_1 + X_2) = \gamma(X_1 + X_2),$$

where $\mathbb{R} \ni \gamma \neq 0$ is a non-random constant, then X_1 (and X_2) is an affine transformation of a random variable with the Poisson distribution (ie. X_1 has the displaced Poisson type distribution).

Problem 7.9 *Prove the following variant of Theorem 7.1.2:*

If X_1, X_2 are i. i. d. random variables with finite second moments, and

$$Var(X_1 - X_2 | X_1 + X_2) = \gamma(X_1 + X_2)^2,$$

where $\mathbb{R} \ni \gamma > 0$ is a non-random constant, then X_1 (and X_2) is an affine transformation of a random variable with the gamma distribution (ie. X_1 has displaced gamma type distribution).

Chapter 8

Gaussian processes

In this chapter we shall consider characterization questions for stochastic processes. We shall treat a stochastic process $\mathbf{X}$ as a function $X_t(\omega)$ of two arguments $t \in [0,1]$ and $\omega \in \Omega$ that are measurable in argument ω, ie. as an uncountable family of random variables $\{X_t\}_{0 \le t \le 1}$. We shall also encounter processes with continuous trajectories, that is processes where functions $X_t(\omega)$ depend continuously on argument t (except on a set of ω's of probability 0).

8.1　Construction of the Wiener process

The Wiener process was constructed and analyzed by Norbert Wiener [150] (please note the date). In the literature, the Wiener process is also called the *Brownian motion,* for Robert Brown, who frequently (and apparently erroneously) is credited with the first observations of chaotic motions in suspension; Nelson [115] gives an interesting historical introduction and lists relevant works prior to Brown. Since there are other more exact mathematical models of the Brownian motion available in the literature, cf. Nelson [115] (see also [17]), we shall stick to the above terminology. The reader should be however aware that in probabilistic literature Wiener's name is nowadays more often used for the measure on the space $C[0,1]$, generated by what we shall call the Wiener process.

The simplest way to define the Wiener process is to list its properties as follows.

Definition 8.1.1 *The Wiener process* $\{W_t\}$ *is a Gaussian process with continuous trajectories such that*

$$
\begin{aligned}
W_0 &= 0; & (8.1) \\
EW_t &= 0 \ \textit{for all } t \ge 0; & (8.2) \\
EW_t W_s &= \min\{t,s\} \ \textit{for all } t,s \ge 0. & (8.3)
\end{aligned}
$$

Recall that a stochastic process $\{X_t\}_{0 \le t \le 1}$ is Gaussian, if the n-dimensional r. v. $(X_{t_1}, \ldots, X_{t_n})$ has multivariate normal distribution for all $n \ge 1$ and all $t_1, \ldots, t_n \in [0,1]$. A stochastic process $\{X_t\}_{t \in [0,1]}$ has continuous trajectories if it is defined by a $C[0,1]$-valued random vector, cf. Example 3.2.2. For infinite time interval $t \in [0, \infty)$, a stochastic process has continuous trajectories if its restriction to $t \in [0, N]$ has continuous trajectories for all $N \in \mathbb{N}$.

The definition of the Wiener process lists its important properties. In particular, conditions (8.1)–(8.3) imply that the Wiener process has independent increments, ie. $W_0, W_t - W_0, W_{t+s} - W_t, \ldots$ are independent. The definition has also one obvious deficiency; it does not say whether a process with all the required properties does exist (the Kolmogorov Existence Theorem [9, Theorem 36.1] does not guarantee continuity of the trajectories.) In this section we answer the existence question by an analytical proof which matches well complex analysis methods used in this book; for a couple of other constructions, see Ito & McKean [66].

The first step of construction is to define an appropriate Gaussian random variable W_t for each fixed t. This is accomplished with the help of the series expansion (8.4) below. It might be worth emphasizing that every Gaussian process X_t with continuous trajectories has a series representation of a form $X(t) = f_0(t) + \sum \gamma_k f_k(t)$, where $\{\gamma_k\}$ are i. i. d. normal $N(0,1)$ and f_k are deterministic continuous functions. Theorem 2.2.5 is a finite dimensional variant of this expansion. Series expansion questions in more abstract setup are studied in [24], see also references therein.

Lemma 8.1.1 *Let $\{\gamma_k\}_{k\geq 1}$ be a sequence of i. i. d. normal $N(0,1)$ random variables. Let*

$$W_t = \frac{2}{\pi} \sum_{k=0}^{\infty} \frac{1}{2k+1} \gamma_k \sin(2k+1)\pi t. \tag{8.4}$$

Then series (8.4) converges, $\{W_t\}$ is a Gaussian process and (8.1), (8.2) and (8.3) hold for each $0 \leq t, s \leq \frac{1}{2}$.

Proof. Obviously series (8.4) converges in the L_2 sense (ie. in mean-square), so random variables $\{W_t\}$ are well defined; clearly, each finite collection $W_{t_1}, \ldots, W_{t_k}$ is jointly normal and (8.1), (8.2) hold. The only fact which requires proof is (8.3). To see why it holds, and also how the series (8.4) was produced, for $t, s \geq 0$ write $\min\{t, s\} = \frac{1}{2}(|t + s| - |t - s|)$. For $|x| \leq 1$ expand $f(x) = |x|$ into the Fourier series. Standard calculations give

$$|x| = \frac{1}{2} - \frac{4}{\pi^2} \sum_{k=0}^{\infty} \frac{1}{(2k+1)^2} \cos(2k+1)\pi x. \tag{8.5}$$

Hence by trigonometry

$$\min\{t, s\} = \frac{2}{\pi^2} \sum_{k=0}^{\infty} \frac{1}{(2k+1)^2} \left(\cos\left((2k+1)\pi(t-s)\right) - \cos\left((2k+1)\pi(t+s)\right) \right)$$

$$= \frac{4}{\pi^2} \sum_{k=0}^{\infty} \frac{1}{(2k+1)^2} \sin((2k+1)\pi t) \sin((2k+1)\pi s).$$

From (8.4) it follows that $EW_t W_s$ is given by the same expression and hence (8.3) is proved. $\square$

To show that series (8.4) converges in probability[1] uniformly with respect to t, we need to analyze $\sup_{0 \leq t \leq 1/2} |\sum_{k \geq n} \frac{1}{2k+1} \gamma_k \sin(2k+1)\pi t|$. The next lemma analyzes instead the expression $\sup_{\{z \in \mathbb{C}: |z|=1\}} |\sum_{k \geq n} \frac{1}{2k+1} \gamma_k z^{2k+1}|$, the latter expression being more convenient from the analytic point of view.

---

[1] Notice that this suffices to prove the existence of the Wiener process $\{W_t\}_{0 \leq t \leq 1/2}$!

Lemma 8.1.2 *There is $C > 0$ such that*

$$E\sup_{|z|=1}\left|\sum_{k=m}^{n}\frac{1}{2k+1}\gamma_k z^{2k+1}\right|^4 \le C(n-m)\left(\sum_{k=m}^{n}\frac{1}{(2k+1)^2}\right)^2 \tag{8.6}$$

for all $m, n \ge 1$.

Proof. By Cauchy's integral formula

$$\left(\sum_{k=m}^{n}\frac{1}{2k+1}\gamma_k z^{2k+1}\right)^4 = z^{2m}\left(\sum_{k=0}^{n-m}\frac{1}{2k+2m+1}\gamma_k z^{2k+1}\right)^4$$

$$= z^{2m}\frac{1}{2\pi i}\oint_L\left(\sum_{k=0}^{n-m}\frac{1}{2k+2m+1}\gamma_k \zeta^{2k+1}\right)^4\frac{1}{\zeta-z}d\zeta,$$

where $L \subset \mathbf{C}$ is the circle $|\zeta| = 1 + 1/(n-m)$.

Therefore

$$\sup_{|z|=1}\left|\sum_{k=m}^{n}\frac{1}{2k+1}\gamma_k z^{2k+1}\right|^4$$

$$\le \sup_{|z|=1}\frac{1}{|\zeta-z|}\frac{1}{2\pi}\oint_L\left|\sum_{k=0}^{n-m}\frac{1}{2k+2m+1}\gamma_{k+m}\zeta^{2k+1}\right|^4 d\zeta.$$

Obviously $\sup_{|z|=1}\frac{1}{|\zeta-z|} = n-m$, and furthermore we have $|\zeta^{2k+1}| \le (1+1/(n-m))^{2k+1} \le e^3$ for all $0 \le k \le n-m$ and all $\zeta \in L$. Hence

$$E\sup_{|z|=1}\left|\sum_{k=m}^{n}\frac{1}{2k+1}\gamma_k z^{2k+1}\right|^4 \le C(n-m)\oint_L E\left|\sum_{k=0}^{n-m}\frac{1}{2k+2m+1}\gamma_k \zeta^{2k+1}\right|^4 d\zeta$$

$$\le C_1(n-m)\left(\sum_{k=0}^{n-m}\frac{1}{(2k+2m+1)^2}\right)^2,$$

which concludes the proof. $\square$

Now we are ready to show that the Wiener process exists.

Theorem 8.1.3 *There is a Gaussian process $\{W_t\}_{0\le t\le 1/2}$ with continuous trajectories and such that (8.1), (8.2), and (8.3) hold.*

Proof. Let W_t be defined by (8.4). By Lemma 8.1.1, properties (8.1)–(8.3) are satisfied and $\{W_t\}$ is Gaussian. It remains to show that series $\sum_{k=0}^{\infty}\frac{1}{2k+1}\gamma_k \sin(2k+1)\pi t$ converges in probability with respect to the supremum norm in $C[0,\frac{1}{2}]$. Indeed, each term of this series is a $C[0,\frac{1}{2}]$-valued random variable and limit in probability defines $\{W_t\}_{0\le t\le 1/2} \in C[0,\frac{1}{2}]$ on the set of ω's of probability one. We need therefore to show that for each $\epsilon > 0$

$$P\Big(\sup_{0\le t\le 1/2}\Big|\sum_{k=n}^{\infty}\frac{1}{2k+1}\gamma_k \sin(2k+1)\pi t\Big| > \epsilon\Big) \to 0 \text{ as } n \to \infty.$$

Since $\sin(2k+1)\pi t$ is the imaginary part of z^{2k+1} with $z = e^{i\pi t}$, it suffices to show that $P(\sup_{|z|=1} \left| \sum_{k=n}^{\infty} \frac{1}{2k+1} \gamma_k z^{2k+1} \right| > \epsilon) \to 0$ as $n \to \infty$. Let r be such that $2^{r-1} < n \leq 2^r$. Notice that by triangle inequality (for the L_4-norm) we have

$$\left(E \sup_{|z|=1} \left| \sum_{k=n}^{\infty} \frac{1}{2k+1} \gamma_k z^{2k+1} \right|^4 \right)^{1/4}$$

$$\leq \left(E \sup_{|z|=1} \left| \sum_{k=n}^{2^r} \frac{1}{2k+1} \gamma_k z^{2k+1} \right|^4 \right)^{1/4}$$

$$+ \sum_{j=r}^{\infty} \left(E \sup_{|z|=1} \left| \sum_{k=2^j+1}^{2^{j+1}} \frac{1}{2k+1} \gamma_k z^{2k+1} \right|^4 \right)^{1/4} .$$

From (8.6) we get

$$\left(E \sup_{|z|=1} \left| \sum_{k=n}^{2^r} \frac{1}{2k+1} \gamma_k z^{2k+1} \right|^4 \right)^{1/4} \leq C 2^{-r/4},$$

and similarly

$$\left(E \left\{ \sup_{|z|=1} \left| \sum_{k=2^j+1}^{2^{j+1}} \frac{1}{2k+1} \gamma_k z^{2k+1} \right|^4 \right\} \right)^{1/4} \leq C 2^{-j/4}$$

for every $j \geq r$. Therefore

$$\left(E \sup_{|z|=1} \left| \sum_{k=n}^{\infty} \frac{1}{2k+1} \gamma_k z^{2k+1} \right|^4 \right)^{1/4} \leq C 2^{-r/4} + \sum_{j=r+1}^{\infty} C 2^{-j/4} \leq C n^{-1/4} \to 0$$

as $n \to \infty$ and convergence in probability (in the uniform metric) follows from Chebyshev's inequality. $\square$

Remark: Usually the Wiener process is considered on unbounded time interval $[0, \infty)$. One way of constructing such a process is to *glue in* countably many independent copies $W, W', W'', \ldots$ of the Wiener process $\{W_t\}_{0 \leq t \leq 1/2}$ constructed above. That is put

$$\widetilde{W}_t = \begin{cases} W_t & \text{for } 0 \leq t \leq \frac{1}{2}, \\ W_{1/2} + W'_{t-1/2} & \text{for } \frac{1}{2} \leq t \leq 1, \\ W_{1/2} + W'_1 + W''_{t-1/2} & \text{for } 1 \leq t \leq \frac{3}{2}, \\ \vdots \end{cases}$$

Since each copy $W^{(k)}$ starts at 0, this construction preserves the continuity of trajectories, and the increments of the resulting process are still independent and normal.

8.2   Levy's characterization theorem

In this section we shall characterize Wiener process by the properties of the first two conditional moments. We shall use conditioning with respect to the past σ-field $\mathcal{F}_s = \sigma\{X_t : t \leq s\}$ of a stochastic process $\{X_t\}$. The result is due to P. Levy [98, Theorem 67.3]. Dozzi [40, page 147 Theorem 1] gives a related multi-parameter result.

Theorem 8.2.1 *If a stochastic process $\{X_t\}_{0 \le t \le 1}$ has continuous trajectories, is square integrable, and*

$$E\{X_t | \mathcal{F}_s\} \;\; = X_s \text{ for all } s \le t; \tag{8.7}$$

$$Var(X_t | \mathcal{F}_s) \;\; = t - s \text{ for all } s \le t, \tag{8.8}$$

then $\{X_t\}$ is the Wiener process.

Conditions (8.7) and (8.8) resemble assumptions made in Chapter 7, cf. Theorems 7.2.1 and 7.5.1. Clearly, formulas (8.7) and (8.8) hold also true for the Poisson process; hence the assumption of continuity of trajectories is essential. The actual role of the continuity assumption is hardly visible, until a stochastic integrals approach is adopted (see, eg. [41, Section 2.11]); then it becomes fairly clear that the continuity of trajectories allows insights into the *future* of the process (compare also Theorem 7.5.1; the latter can be thought as a discrete-time analogue of Levy's theorem.). Neveu [117, Ch. 7] proves several other discrete time versions of Theorem 8.2.1 that are of different nature.

Proof of Theorem 8.2.1. Let $0 \le s \le 1$ be fixed. Put $\phi(t, u) = E\{\exp(iuX_{t+s}) | \mathcal{F}_s\}$. Clearly $\phi(\cdot, \cdot)$ is continuous with respect to both arguments. We shall show that

$$\frac{\partial}{\partial t}\phi(t, u) = -\frac{1}{2}u^2 \phi(t, u) \tag{8.9}$$

almost surely (with the derivative defined with respect to convergence in probability). This will conclude the proof, since equation (8.9) implies

$$\phi(t, u) = \phi(0, u)e^{-tu^2/2} \tag{8.10}$$

almost surely. Indeed, (8.10) means that the increments $X_{t+s} - X_s$ are independent of the past $\mathcal{F}_s$ and have normal distribution with mean 0 and variance t, ie. $\{X_t\}$ is a Gaussian process, and (8.1)–(8.3) are satisfied.

It remains to verify (8.9). We shall consider the right-hand side derivative only; the left-hand side derivative can be treated similarly and the proof shows also that the derivative exists. Since u is fixed, through the argument below we write $\phi(t) = \phi(t, u)$. Clearly

$$\phi(t + h) - \phi(t) = E\{\exp(iuX_t)(e^{iu(X_{t+h} - X_t)} - 1) | \mathcal{F}_s\}$$

$$= -\frac{1}{2}u^2 h \phi(t) + E\{\exp(iuX_t)R(X_{t+h} - X_t) | \mathcal{F}_s\},$$

where $|R(x)| \le |x|^3$ is the remainder in Taylor's expansion for e^x. The proof will be concluded, once we show that $E\{|X_{t+h} - X_t|^3 | \mathcal{F}_s\}/h \to 0$ as $h \to 0$. Moreover, since we require convergence in probability, we need only to verify that $E|X_{t+h} - X_t|^3/h \to 0$. It remains therefore to establish the following lemma, taken (together with the proof) from [7, page 25, Lemma 3.2]. $\square$

Lemma 8.2.2 *Under the assumptions of Theorem 8.2.1 we have*

$$E|X_{t+h} - X_t|^4 < \infty.$$

Moreover, there is $C > 0$ such that

$$E|X_{t+h} - X_t|^4 \le Ch^2$$

for all $t, h \ge 0$.

Proof. We discretize the interval $(t, t+h)$ and write $Y_k = X_{t+kh/N} - X_{t+(k-1)h/N}$, where $1 \le k \le N$. Then

$$|X_{t+h} - X_t|^4 = \sum_k Y_k^4 \tag{8.11}$$
$$+ \ 4 \sum_{m \neq n} Y_m^3 Y_n + 3 \sum_{m \neq n} Y_m^2 Y_n^2$$
$$+ \ 6 \sum_{k \neq m \neq n} Y_m^2 Y_n Y_k + \sum_{k \neq l \neq m \neq n} Y_k Y_l Y_m Y_n.$$

Using elementary inequality $2ab \le a^2/\theta + b^2\theta$, where $\theta > 0$ is arbitrary, we get

$$\sum_k Y_k^4 + 4 \sum_{m \neq n} Y_m^3 Y_n \tag{8.12}$$
$$= \ 4 \sum_n Y_n^3 \sum_m Y_m - 3 \sum_n Y_n^4 \le 2\theta^{-1} (\sum_n Y_n^3)^2 + 2\theta (\sum_m Y_m)^2.$$

Notice, that

$$|\sum_n Y_n^3| \to 0 \tag{8.13}$$

in probability as $N \to \infty$. Indeed, $|\sum_n Y_n^3| \le \sum_n Y_n^2 |Y_n| \le \max_n |Y_n| \sum_n Y_n^2$. Therefore for every $\epsilon > 0$

$$P(|\sum_n Y_n^3| > \epsilon) \le P(\sum_n Y_n^2 > M) + P(\max_n |Y_n| > \epsilon/M).$$

By (8.8) and Chebyshev's inequality $P(\sum_n Y_n^2 > M) \le h/M$ is arbitrarily small for all M large enough. The continuity of trajectories of $\{X_t\}$ implies that for each M we have $P(\max_n |Y_n| > \epsilon/M) \to 0$ as $N \to \infty$, which proves (8.13).

Passing to a subsequence, we may therefore assume $|\sum_n Y_n^3| \to 0$ as $N \to \infty$ with probability one. Using Fatou's lemma (see eg. [9, Ch. 3 Theorem 16.3]), by continuity of trajectories and (8.11), (8.12) we have now

$$E\left\{|X_{t+h} - X_t|^4\right\} \le \limsup_{N \to \infty} E\left\{ 2\theta (\sum_m Y_m)^2 + 3 \sum_{m \neq n} Y_m^2 Y_n^2 \right. \tag{8.14}$$
$$\left. + 6 \sum_{k \neq m \neq n} Y_m^2 Y_n Y_k + \sum_{k \neq l \neq m \neq n} Y_k Y_l Y_m Y_n \right\},$$

provided the right hand side of (8.14) is integrable.

We now show that each term on the right hand side of (8.14) is integrable and give the bounds needed to conclude the argument. The first two terms are handled as follows.

$$E(\sum_m Y_m)^2 \le h; \tag{8.15}$$
$$EY_m^2 Y_n^2 = \lim_{M \to \infty} EY_m^2 I_{|Y_m| \le M} Y_n^2 \tag{8.16}$$
$$= \lim_{M \to \infty} EY_m^2 I_{|Y_m| \le M} E\{Y_n^2 | \mathcal{F}_{t+hm/N}\} \le h^2/N^2 \text{ for all } m < n;$$

Considering separately each of the following cases: $m < n < k, m < k < n, n < m < k,$ $n < k < m, k < m < n, k < n < m$, we get $E|Y_m^2 Y_n Y_k| \leq h^2/N^2 < \infty$. For instance, the case $m < n < k$ is handled as follows

$$E|Y_m^2 Y_n Y_k| = \lim_{M \to \infty} E\left\{Y_m^2 |Y_n| I_{|Y_m| \leq M} E\left\{|Y_k| \,\big|\, \mathcal{F}_{t+hm/N}\right\}\right\}$$

$$\leq \lim_{M \to \infty} E\left\{Y_m^2 |Y_n| I_{|Y_m| \leq M} (E\{Y_k^2 \mathcal{F}_{t+hm/N}\})^{1/2}\right\}$$

$$= (h/N)^{1/2} \lim_{M \to \infty} E\left\{Y_m^2 \, I_{|Y_m| \leq M} E\left\{|Y_n| \,\big|\, \mathcal{F}_{t+hk/N}\right\}\right\}$$

$$\leq (h/N)^{1/2} \lim_{M \to \infty} E\left\{Y_m^2 I_{|Y_m| \leq M} (E\{Y_n^2 | \mathcal{F}_{t+hk/N}\})^{1/2}\right\} = h^2/N^2.$$

Once $E|Y_m^2 Y_n Y_k| < \infty$ is established, it is trivial to see from (8.7) in each of the cases $m < n < k, m < k < n, n < m < k, k < m < n$, (and using in addition (8.8) in the cases $n < k < m, k < n < m$) that

$$EY_m^2 Y_n Y_k = 0 \tag{8.17}$$

for every choice of different numbers m, n, k. Analogous considerations give $E|Y_m Y_n Y_k Y_l| \leq h^2/N^2 < \infty$. Indeed, suppose for instance that $m < k < l < n$. Then

$$E|Y_m Y_n Y_k Y_l|$$

$$= \lim_{M \to \infty} E\{|Y_m| I_{|Y_m| \leq M} |Y_n| I_{|Y_n| \leq M} |Y_k| I_{|Y_k| \leq M} E\left\{|Y_l| \,\big|\, \mathcal{F}_{t+hk/N}\right\}\}$$

$$\leq \lim_{M \to \infty} E\left\{|Y_m| I_{|Y_m| \leq M} |Y_n| I_{|Y_n| \leq M} |Y_k| I_{|Y_k| \leq M} (E\{Y_l^2 | \mathcal{F}_{t+hk/N}\})^{1/2}\right\}$$

$$= (h/N)^{1/2} E|Y_m Y_n Y_k|,$$

and the procedure continues replacing one variable at a time by the factor $(h/N)^{1/2}$. Once $E|Y_m Y_n Y_k Y_l| < \infty$ is established, (8.7) gives trivially

$$EY_m Y_n Y_k Y_l = 0 \tag{8.18}$$

for every choice of different m, n, k, l. Then (8.15)–(8.18) applied to the right hand side of (8.14) give $E|X_{t+h} - X_t|^4 \leq 2\theta h + 3h^2$. Since θ is arbitrarily close to 0, this ends the proof of the lemma. $\square$

The next result is a special case of the theorem due to J. Jakubowski & S. Kwapień [67]. It has interesting applications to convergence of random series questions, see [91] and it also implies Azuma's inequality for martingale differences.

Theorem 8.2.3 *Suppose $\{X_k\}$ satisfies the following conditions*
(i) $|X_k| \leq 1, k = 1, 2, \ldots$;
(ii) $E\{X_{n+1}|X_1, \ldots, X_n\} = 0, n = 1, 2, \ldots$.
Then there is an i. i. d. symmetric random sequence $\epsilon_k = \pm 1$ and a σ-field $\mathcal{N}$ such that the sequence

$$Y_k = E\{\epsilon_k | \mathcal{N}\}$$

has the same joint distribution as $\{X_k\}$.

Proof. We shall first prove the theorem for a finite sequence $\{X_k\}_{k=1,\ldots n}$. Let $F(dy_1,\ldots,dy_n)$ be the joint distribution of $X_1,\ldots,X_n$ and let $G(du) = \frac{1}{2}(\delta_{-1}+\delta_1)$ be the distribution of ϵ_1. Let $P(dy,du)$ be a probability measure on $\mathbb{R}^{2n}$, defined by

$$P(dy,du) = \prod_{j=1}^{n}(1+u_jy_j)F(dy_1,\ldots,dy_n)G(du_1)\ldots G(du_n) \qquad (8.19)$$

and let $\mathcal{N}$ be the σ-field generated by the y-coordinate in $\mathbb{R}^{2n}$. In other words, take the joint distribution Q of independent copies of (X_k) and ϵ_k and define P on $\mathbb{R}^{2n}$ as being absolutely continuous with respect to Q with the density $\prod_{j=1}^{n}(1+u_jy_j)$. Using Fubini's theorem (the integrand is non-negative) it is easy to check now, that $P(dy,\mathbb{R}^n) = F(dy)$ and $P(\mathbb{R}^n,du) = G(du_1)\ldots G(du_n)$. Furthermore $\int u_j \prod_{j=1}^{n}(1+u_jy_j)G(du_1)\ldots G(du_n) = y_j$ for all j, so the representation $E\{\epsilon_j|\mathcal{N}\} = Y_j$ holds. This proves the theorem in the case of the finite sequence $\{X_k\}$.

To construct a probability measure on $\mathbb{R}^{\infty} \times \mathbb{R}^{\infty}$, pass to the limit as $n \to \infty$ with the measures P_n constructed in the first part of the proof; here P_n is treated as a measure on $\mathbb{R}^{\infty} \times \mathbb{R}^{\infty}$ which depends on the first $2n$ coordinates only and is given by (8.19). Such a limit exists along a subsequence, because P_n is concentrated on a compact set $[-1,1]^{\mathbb{N}}$ and hence it is tight. $\square$

8.3　Characterizations of processes without continuous trajectories

Recall that a stochastic process $\{X_t\}$ is L_2, or mean-square continuous, if $X_t \in L_2$ for all t and $X_t \to X_{t_0}$ in L_2 as $t \to t_0$, cf. Section 1.2. Similarly, X_t is mean-square differentiable, if $t \mapsto X_t \in L_2$ is differentiable as a Hilbert-space-valued mapping $\mathbb{R} \to L_2$. For mean zero processes, both are the properties[2] of the covariance function $K(t,s) = EX_tX_s$.

Let us first consider a simple result from [18][3], which uses L_2-smoothness of the process, rather than continuity of the trajectories, and uses only conditioning by one variable at a time. The result does not apply to processes with non-smooth covariance, such as (8.3).

Theorem 8.3.1 *Let $\{X_t\}$ be a square integrable, L_2-differentiable process such that for every $t \geq 0$ the correlation coefficient between random variables X_t and $\frac{d}{dt}X_t$ is strictly between -1 and 1. Suppose furthermore that*

$$E\{X_t|X_s\} \quad \text{is a linear function of } X_s \text{ for all } s < t; \qquad (8.20)$$

$$Var(X_t|X_s) \qquad \text{is non-random for all } s < t. \qquad (8.21)$$

Then the one dimensional distributions of $\{X_t\}$ are normal.

---

[2] For instance, $E(X_t - X_s)^2 = K(t,t) + K(s,s) - 2K(t,s)$, so mean square continuity follows from the continuity of the covariance function $K(t,s)$.

[3] See also [139].

Lemma 8.3.2 *Let X, Y be square integrable standardized random variables such that $\rho = EXY \neq \pm 1$. Assume $E\{X|Y\} = \rho Y$ and $Var(X|Y) = 1 - \rho^2$ and suppose there is an L_2-differentiable process $\{Z_t\}$ such that*

$$Z_0 = Y; \tag{8.22}$$

$$\frac{d}{dt}Z_t\big|_{t=0} = X. \tag{8.23}$$

Furthermore, suppose that

$$E\{X|Z_t\} - a_t Z_t \to 0 \text{ in } L_2 \text{ as } t \to 0, \tag{8.24}$$

where $a_t = corr(X, Z_t)/Var(Z_t)$ is the linear regression coefficient. Then Y is normal.

Proof. It is straightforward to check that $a_0 = \rho$ and $\frac{d}{dt}a_t\big|_{t=0} = 1 - 2\rho^2$. Put $\phi(t) = E\exp(itY)$ and let $\psi(t, s) = EZ_s \exp(itZ_s)$. Clearly $\psi(t, 0) = -i\frac{d}{dt}\phi(t)$. These identities will be used below without further mention. Put $V_s = E\{X|Z_s\} - a_s Z_s$. Trivially, we have

$$E\{X \exp(itZ_s)\} = a_s \psi(t, s) + E\{V_s \exp(itZ_s)\}. \tag{8.25}$$

Notice that by the L_2-differentiability assumption, both sides of (8.25) are differentiable with respect to s at $s = 0$. Since by assumption $V_0 = 0$ and $V_0' = 0$, differentiating (8.25) we get

$$itE\{X^2 \exp(itY)\}$$
$$= (1 - 2\rho^2)\psi(t, 0) + \rho E\{X \exp(itY)\} \quad + \quad it\rho E\{XY \exp(itY)\}. \tag{8.26}$$

Conditional moment formulas imply that

$$E\{X \exp(itY)\} = \rho E\{Y \exp(itY)\} = -i\rho\phi'(t)$$

$$E\{XY \exp(itY)\} = \rho E\{Y^2 \exp(itY)\} = -\rho^2 \phi''(t)$$

$$E\{X^2 \exp(itY)\} = (1 - \rho^2)\phi(t) + \rho^2 \phi''(t),$$

see Theorem 1.5.3. Plugging those relations into (8.26) we get

$$(1 - \rho^2)it\phi(t) = -(1 - \rho^2)i\phi'(t),$$

which, since $\rho^2 \neq 1$, implies $\phi(t) = e^{-t^2/2}$. $\square$

Proof of Theorem 8.3.1. For each fixed $t_0 > 0$ apply Lemma 8.3.2 to random variable $X = \frac{d}{dt}X_{t_0}, Y = X_t$ with $Z_t = X_{t_0-t}$. The only assumption of Lemma 8.3.2 that needs verification is that $Var(X|Y)$ is non-random. This holds true, because in L_1-convergence

$$Var(X|Y) = \lim_{\epsilon \to 0} \epsilon^{-1} E\{(X_{t_0+\epsilon} - Y - \rho(\epsilon)Y)^2 | Y\}$$

$$= \lim_{\epsilon \to 0} \epsilon^{-1} Var(X_{t_0+\epsilon}|X_{t_0}) = \rho'(0),$$

where $\rho(h) = E\{X_{t_0+h}X_{t_0}\}/EX_{t_0}^2$. Therefore by Lemma 8.3.2, X_t is normal for all $t > 0$. Since X_0 is the L_2-limit of X_t as $t \to 0$, hence X_0 is normal, too. $\square$

As we pointed out earlier, Theorem 8.2.1 is not true for processes without continuous trajectories. In the next theorem we use σ-fields that allow some *insight into the future* rather than past σ-fields $\mathcal{F}_s = \sigma\{X_t : t \leq s\}$. Namely, put

$$\mathcal{G}_{s,u} = \sigma\{X_t : t \leq s \text{ or } t = u\}$$

The result, originally under minor additional technical assumptions, comes from [121]. The proof given below follows [19].

Theorem 8.3.3 *Suppose $\{X_t\}_{0\leq t\leq 1}$ is an L_2-continuous process such that $corr(X_t, X_s) \neq \pm 1$ for all $t \neq s$. If there are functions $a(s,t,u)$, $b(s,t,u)$, $c(s,t,u)$, $\sigma^2(s,t,u)$ such that for every choice of $s \leq t$ and every u we have*

$$\begin{aligned}
E\{X_t|\mathcal{G}_{s,u}\} &= a(s,t,u) + b(s,t,u)X_s + c(s,t,u)X_u; & (8.27) \\
Var(X_t|\mathcal{G}_{s,u}) &= \sigma^2(s,t,u), & (8.28)
\end{aligned}$$

then $\{X_t\}$ is Gaussian.

The proof is based on the following version of Lemma 7.5.2.

Lemma 8.3.4 *Let $N \geq 1$ be fixed and suppose that $\{X_n\}$ is a sequence of square integrable random variables such that the following conditions, compare (7.18)–(7.21), hold for all $n \geq 1$:*

$$\begin{aligned}
E\{X_{n+1}|X_1,\ldots,X_n\} &= c_1 X_n + c_2, \\
E\{X_{n+1}|X_1,\ldots,X_n,X_{n+2}\} &= c_3 X_n + c_4 X_{n+2} + c_5, \\
Var(X_{n+1}|X_1,\ldots,X_n) &= c_6, \\
Var(X_{n+1}|X_1,\ldots,X_n,X_{n+2}) &= c_7.
\end{aligned}$$

Moreover, suppose that the correlation coefficient $\rho_n = corr(X_n, X_{n+1})$ satisfies $\rho_n^2 \neq 0, 1$ for all $n \geq N$. If $(X_1,\ldots,X_{N-1})$ is jointly normal, then $\{X_k\}$ is Gaussian.

If $N = 1$, Lemma 8.3.4 is the same as Lemma 7.5.4; the general case $N \geq 1$ is proved similarly, except that since $(X_1,\ldots,X_{N-1})$ is normal, one needs to calculate conditional moments $x = E\{(X_{n+1} - \rho^2 X_{n-1})^k|X_1,\ldots,X_{n-1}\}$ and $y = E\{(X_n - \rho X_{n-1})^k|X_1,\ldots,X_{n-1}\}$ for $n \geq N$ only. Also, not assuming Markov property here, one needs to consider the above expressions which are based on conditioning with respect to past σ-field, rather than (7.22) and (7.23). The detailed proof can be found in [19].

Proof of Theorem 8.3.3. Let $\{t_n\}$ be the sequence running through the set of all rational numbers. We shall show by induction that for each $N \geq 1$ random sequence $(X_{t_1}, X_{t_2},\ldots,X_{t_N})$ has the multivariate normal distribution. Since $\{t_n\}$ is dense and X_t is L_2-continuous, this will prove that $\{X_t\}$ is a Gaussian process.

To proceed with the induction, suppose $(X_{t_1}, X_{t_2},\ldots,X_{t_{N-1}})$ is normal for some $N \geq 1$ (with the convention that the empty set of random variables is normal). Let

$s_1 < s_2 < \ldots$ be an infinite sequence such that $\{s_1, \ldots, s_N\} = \{t_1, \ldots, t_N\}$ and furthermore $corr(X_{s_k}, X_{s_{k+1}}) \neq 0$ for all $k \geq N$. Such a sequence exists by L_2-continuity; given $s_1, \ldots, s_k$, we have $EX_{s_k} X_s \to EX_{s_k}^2$ as $s \downarrow s_k$, so that an appropriate rational $s_{k+1} \neq s_k$ can be found. Put $X_n = X_{s_n}, n \geq 1$. Then the assumptions of Lemma 8.3.4 are satisfied: correlation coefficients are not equal ± 1 because s_k are different numbers; conditional moment assumption holds by picking the appropriate values of t, u in (6.3.8) and (6.3.9). Therefore Lemma 6.3.5 implies that $(X_{t_1}, \ldots, X_{t_N})$ is normal and by induction the proof is concluded.$\Box$

Remark: A variant of Theorem 8.3.3 for the Wiener process obtained by specifying suitable functions $a(s, t, u), b(s, t, u), c(s, t, u), \sigma^2(s, t, u)$ can be deduced directly from Theorem 8.2.1 and Theorem 6.2.2. Indeed, a more careful examination of the proof of Theorem 6.2.2 shows that one gets estimates for $E|X_t - X_s|^4$ in terms of $E|X_t - X_s|^2$. Therefore, by the well known Kolmogorov's criterion ([42, Exercise 1.3 on page 337]) the process has a version with continuous trajectories and Theorem 8.2.1 applies.

The proof given in the text characterizes more general Gaussian processes. It can also be used with minor modifications to characterize other stochastic processes, for instance for the Poisson process, see [21, 147].

8.4 Second order conditional structure

The results presented in Sections 4.1, 7.5, 8.2, and 8.3 suggest the general problem of analyzing what one might call random fields with linear conditional structure. The setup is as follows. Let $(T, \mathcal{B}_T)$ be a measurable space. Consider random field $\mathbf{X} : T \times \Omega \to \mathbb{R}$, where Ω is a probability space. We shall think of $\mathbf{X}$ as defined on probability space $\Omega = T^{\mathbb{R}}$ by $\mathbf{X}(t, \omega) = \omega(t)$. Different random fields then correspond to different assignments of the probability measure P on Ω. For each $t \in T$ let $\mathcal{S}_t$ be a given collection of measurable subsets $F \in \sigma\{X_s : s \neq t\}$. For technical reasons, it is convenient to have $\mathcal{S}_t$ consisting of sets F that depend on a finite number of coordinates only. Even if $T = \mathbb{R}$, the choice of $\mathcal{S}_t$ might differ from the usual choice of the theory of stochastic processes, where $\mathcal{S}_t$ usually consists of those $F \ni s$ with $s < t$.

One can say that $\mathbf{X}$ has linear conditional structure if

Condition 8.4.1 *For each $t \in T$ and every $F \in \mathcal{S}_t$ there is a measure $\alpha(.) = \alpha_{t,F}(.)$ and a number $b = b(t, F)$ such that $E\{X(t)|X(s) : s \in F\} = b + \int_T X(s)\alpha(ds)$.*

Clearly, this definition encompasses many of the examples that were considered in previous sections. When T is a measurable space with a measure μ, one may also be interested in variations of the condition 8.4.1. For instance, if $\mathbf{X}$ has μ-square-integrable trajectories, one can consider the following variant.

Condition 8.4.2 *For each $t \in T$ and $F \in \mathcal{S}_t$ there is a number b and a bounded linear operator $A = A_{t,F} : L_2(T, d\mu) \to L_2(F, d\mu)$ such that $E\{X(t)|X(s) : s \in F\} = b + A(\mathbf{X})$.*

In this notation, Condition 8.4.1 corresponds to the integral operator $Af = \int_F f(x)\, d\mu$.

The assumption that second moments are finite permits sometimes to express operators A in terms of the covariance $K(t, s)$ of the random field $\mathbf{X}$. Namely, the "equation" is

$$K(t, s) = A_{t,F}(K(\cdot, s))$$

for all $s \in F$.

The main interest of the conditional moments approach is in additional properties of a random field with linear conditional structure - properties determined by a *higher order conditional structure* which gives additional information about the form of

$$E\{(X(t))^2|X(s) : s \in F\}. \tag{8.29}$$

Perhaps the most natural question here is how to tackle finite sequences of arbitrary length. For instance, one would want to say that if a large collection of N random variables has linear conditional moments and conditional variances that are quadratic polynomials, then for large N the distribution should be close to say, a normal, or Poisson, or, say, Gamma distribution. A review of state of the art is in [142], but much work still needs to be done. Below we present two examples illustrating the fact that the form of the first two conditional moments can (perhaps) be grasped on intuitive level from the physical description of the phenomenon.

Example 8.4.1 (snapshot of a random vibration) *Suppose a long chain of molecules is observed in the fixed moment of time (a snapshot). Let $X(k)$ be the (vertical) displacement of the k-th molecule, see Figure 8.1.*

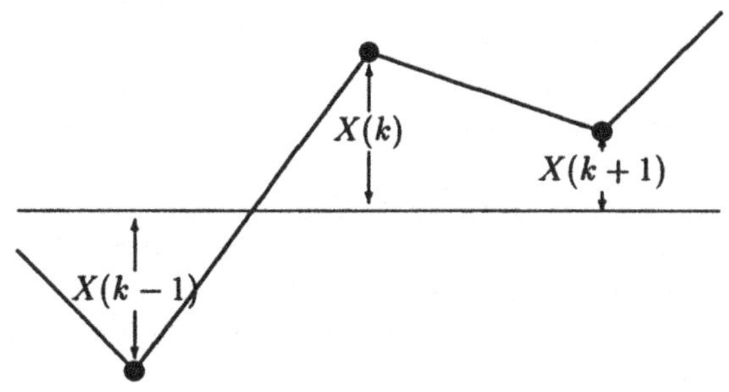

Figure 8.1: Random Molecules

If all positions of the molecules except the k-th one are known, then it is natural to assume that the average position of $X(k)$ is centered between its neighbors, ie.

$$E\{X(k) \quad | \quad given\ all\ other\ positions\ are\ known\} \tag{8.30}$$
$$= \ (X(k-1) + X(k+1))/2.$$

If furthermore we assume that the molecules are connected by elastic springs, then the potential energy of the k-th molecule is proportional to

$$const + (X(k) - X(k-1))^2 + (X(k) - X(k+1))^2 - 1/2(X(k+1) - X(k-1))^2.$$

Therefore, assuming the only source of vibrations is the external heat bath, *the average energy is constant and it is natural to suppose that*

$$E\{(X(k) - X(k-1))^2 + (X(k) - X(k+1))^2$$

$$-1/2(X(k+1) - X(k-1))^2|all\ except\ k\text{-}th\ known\} = const.$$

Using (8.30) this leads after simple calculation to

$$Var(X(k)|\ldots, X(1), \ldots, X(k-1), X(k+1), \ldots) = const, \tag{8.31}$$

and shows to what extend (8.29) might be considered to be "intuitive". To see what might follow from similar conditions, consult [149, Theorem 1] and [147, Theorem 3.1], where various possibilities under quadratic expression (8.29) are listed; to avoid finiteness of all moments, see the proof of [148, Theorem 1.1]. Wesolowski's method for treatment of moments resembles the proof of Lemma 8.2.2; in general it seems to work under broader set of assumptions than the method used in the proof of Theorem 6.2.2.

Example 8.4.2 (a snapshot of epidemic)

Suppose that we observe the development of a disease in a two-dimensional region which was partitioned into many small sub-regions, indexed by a parameter a. Let X_a be the number of infected individuals in the a-th sub-region at the fixed moment of time (a snapshot). If the disease has already spread throughout the whole region, and if in all but the a-th sub-region the situation is known, then we should expect in the a-th sub-region to have

$$E\{X_a|all\ other\ known\} = 1/8 \sum_{b \in neighb(a)} X_b.$$

Furthermore there are some obvious choices for the second order conditional structure, depending on the source of infection: If we have uniform external virus rain, *then*

$$Var(X_a|all\ other\ known) = const. \tag{8.32}$$

On the other hand, if the infection comes from the nearest neighbors only, then, intuitively, the number of infected individuals in the a-th region should be a binomial r. v. with the number of viruses in the neighboring regions as the number of trials. Therefore it is quite natural to assume that

$$Var(X_a|all\ other\ known) = const \sum_{b \in neighb(a)} X_b. \tag{8.33}$$

Clearly, there are many other interesting variants of this model. The simplest would take into account some boundary conditions, *and also perhaps would mix both the virus rain and the infection from the (not necessarily nearest) neighbors. More complicated models could in addition describe the time development of epidemic; for finite periods of time, this amounts to adding another coordinate to the index set of the random field.*

Appendix A

Solutions of selected problems

A.1 Solutions for Chapter 1

Problem 1.1 ([64]) *Hint:* decompose the integral into four terms corresponding to all possible combinations of signs of X, Y. For $X > 0$ and $Y > 0$ use the bivariate analogue of (1.2): $EXY = \int_0^\infty \int_0^\infty P(X > t, Y > s)\,dt\,ds$. Also use elementary identities

$$P(X \geq t, Y \geq s) - P(X \geq t)P(Y \geq s) = P(X \leq t, Y \leq s) - P(X \leq t)P(Y \leq s)$$

$$= -(P(X \leq t, Y \geq s) - P(X \leq t)P(Y \geq s))$$

$$= -(P(X \geq t, Y \leq s) - P(X \geq t)P(Y \leq s)).$$

Problem 1.2 We prove a slightly more general tail condition for integrability, see Corollary 1.3.3.

Claim A.1.1 *Let $X \geq 0$ be a random variable and suppose that there is $C < \infty$ such that for every $0 < \rho < 1$ there is $T = T(\rho)$ such that*

$$P(X > Ct) \leq \rho P(X > t) \text{ for all } t > T. \tag{A.1}$$

Then all the moments of X are finite.

Proof. Clearly, for unbounded random variables (A.1) cannot hold, unless $C > 1$ (and there is nothing to prove if X is bounded). We shall show that inequality (A.1) implies that for $\beta = -\log_C(\rho)$, there are constants $K, T < \infty$ such that

$$N(x) \leq Kx^{-\beta} \text{ for all } x \geq T. \tag{A.2}$$

Since ρ is arbitrarily close to 0, this will conclude the proof , eg. by using formula (1.2).

To prove that (A.1) implies (A.2), put $a_n = C^n T, n = 0, 2, \ldots$. Inequality (A.1) implies

$$N(a_{n+1}) \leq \rho N(a_n), n = 0, 1, 2, \ldots. \tag{A.3}$$

From (A.3) it follows that $N(a_n) \leq N(T)\rho^n$, ie.

$$N(C^{n+1}T) \leq N(T)\rho^n \text{ for all } n \geq 1. \tag{A.4}$$

To end the proof, it remains to observe that for every $x > 0$, choosing n such that $C^n T \le x < C^{n+1} T$, we obtain $N(x) \le N(C^n T) \le C_1 \rho^n$. This proves (A.2) with $K = N(T)\rho^{-1}T^{-\log_C \rho}$. $\square$

Problem 1.3 This is an easier version of Theorem 1.3.1 and it has a slightly shorter proof. Pick $t_0 \ne 0$ and q such that $P(X \ge t_0) < q < 1$. Then $P(|X| \ge 2^n t_0) \le q^{2^n}$ holds for $n = 1$. Hence by induction $P(|X| \ge 2^n t_0) \le q^{2^n}$ for all $n \ge 1$. If $2^n t_0 \le t < 2^{n+1}t_0$, then $P(|X| \ge t) \le P(|X| \ge 2^n t_0) \le q^{2^n} \le q^{t/(2t_0)} = e^{-\theta t}$ for some $\theta > 0$. This implies $E\exp(\lambda|X|) < \infty$ for all $\lambda < \theta$, see (1.2).

Problem 1.4 See the proof of Lemma 2.5.1.

Problem 1.9 Fix $t > 0$ and let $A \in \mathcal{F}$ be arbitrary. By the definition of conditional expectation $\int_A P(|X| > t|\mathcal{F})\,dP = EI_A I_{|X|>t} \le Et^{-1}|X|I_A I_{|X|>t} \le t^{-1}E|X|I_A$. Now use Lemma 1.4.2.

Problem 1.11 $\int_A U\,dP = \int_A V\,dP$ for all $A = X^{-1}(B)$, where B is a Borel subset of $\mathbb{R}$. Lemma 1.4.2 ends the argument.

Problem 1.12 Since the conditional expectation $E\{\cdot|\mathcal{F}\}$ is a contraction on L_1 (or, to put it simply, Jensen's inequality holds for the convex function $x \mapsto |x|$), therefore $|E\{X|Y\}| = |a|E|Y| \le E|X|$ and similarly $|b|E|X| \le E|Y|$. Hence $|ab|E|X|E|Y| \le E|X|E|Y|$.

Problem 1.13 $E\{Y|X\} = 0$ implies $EXY = 0$. Integrating $YE\{X|Y\} = Y^2$ we get $EY^2 = EXY = 0$.

Problem 1.14 We follow [38, page 314]: Since $\int_{X\ge a}(Y - X)\,dP = 0$ and $\int_{Y>b}(Y - X)\,dP = 0$, we have

$$0 \ge \int_{X\ge a, Y\le a}(Y - X)\,dP = \int_{X\ge a}(Y - X)\,dP - \int_{X\ge a, Y>a}(Y - X)\,dP$$

$$= -\int_{X\ge a, Y>a}(Y - X)\,dP = -\int_{Y>a}(Y - X)\,dP + \int_{X<a, Y>a}(Y - X)\,dP$$

$$= \int_{X<a, Y>a}(Y - X)\,dP \le 0$$

therefore $\int_{X<a, Y>a}(Y - X)\,dP = 0$. The integrand is strictly larger than 0, showing that $P(X < a < Y) = 0$ for all rational a. Therefore $X \ge Y$ a. s. and the reverse inequality follows by symmetry.

Problem 1.15 See the proof of Theorem 1.8.1.

Problem 1.16

a) If X has discrete distribution $P(X = x_j) = p_j$, with ordered values $x_j < x_{j+1}$, then for all $\Delta \ge 0$ small enough we have $\phi(x_k + \Delta) = (x_k + \Delta)\sum_{j\le k} p_j + \sum_{j>k} x_j p_j$. Therefore $\lim_{\Delta\to 0} \frac{\phi(x_k+\Delta)-\phi(x_k)}{\Delta} = P(X \le x_k)$.

b) If X has a continuous probability density function $f(x)$, then $\phi(t) = t\int_{-\infty}^t f(x)\,dx + \int_t^\infty x f(x)\,dx$. Differentiating twice we get $f(x) = \phi''(x)$.

For the general case one can use Problem 1.17 (and the references given below).

Problem 1.17 Note: Function $U_\mu(t) = \int |x - t|\mu(dx)$ is called a (one dimensional) *potential of a measure* μ and a lot about it is known, see eg. [26]; several relevant references follow Theorem 4.2.2; but none of the proofs we know is simple enough to be written here.
Formula $|x - t| = 2\max\{x, t\} - x - t$ relates this problem to Problem 1.16.

Problem 1.18 *Hint:* Calculate the variance of the corresponding distribution.

Note: Theorem 2.5.3 gives another related result.

Problem 1.19 Write $\phi(t,s) = \exp Q(t,s)$. Equality claimed in (i) follows immediately from (1.17) with $m = 1$; (ii) follows by calculation with $m = 2$.

Problem 1.20 See for instance [76].

Problem 1.21 Let g be a bounded continuous function. By uniform integrability (cf. (1.18)) $E(Xg(Y)) = \lim_{n\to\infty} E(X_n g(Y_n))$ and similarly $E(Yg(Y)) = \lim_{n\to\infty} E(Y_n g(Y_n))$. Therefore $EXg(Y) = \rho E(Yg(Y))$ for all bounded continuous g. Approximating indicator functions by continuous g, we get $\int_A X\,dP = \int_A \rho Y\,dP$ for all $A = \{\omega : Y(\omega) \in [a,b]\}$. Since these A generate $\sigma(Y)$, this ends the proof.

A.2 Solutions for Chapter 2

Problem 2.1 Clearly $\phi(t) = e^{-t^2/2} \frac{1}{\sqrt{2\pi}} \int_{-\infty}^{\infty} e^{-(x-it)^2/2}\,dx$. Since $e^{-z^2/2}$ is analytic in complex plane $\mathbb{C}$, the integral does not depend on the path of integration, ie. $\int_{-\infty}^{\infty} e^{-(x-it)^2/2}\,dx = \int_{-\infty}^{\infty} e^{-x^2/2}\,dx$.

Problem 2.2 Suppose for simplicity that the random vectors $\mathbf{X}, \mathbf{Y}$ are centered. The joint characteristic function $\phi(\mathbf{t},\mathbf{s}) = E\exp(i\mathbf{t}\cdot\mathbf{X} + i\mathbf{s}\cdot\mathbf{Y})$ equals $\phi(\mathbf{t},\mathbf{s}) = \exp(-\frac{1}{2}E(\mathbf{t}\cdot\mathbf{X})^2 \exp(-\frac{1}{2}E(\mathbf{s}\cdot \mathbf{Y})^2)\exp(-E(\mathbf{t}\cdot\mathbf{X})(\mathbf{s}\cdot\mathbf{Y}))$. Independence follows, since $E(\mathbf{t}\cdot\mathbf{X})(\mathbf{s}\cdot\mathbf{Y}) = \sum_{i,j} t_i s_j EX_iY_j = 0$.

Problem 2.3 Here is a heavy-handed approach: Integrating (2.9) in polar coordinates we express the probability in question as $\int_0^{\pi/2} \frac{\cos 2\theta}{1 - \sin 2\alpha \sin 2\theta}\,d\alpha$. Denoting $z = e^{2i\theta}, \xi = e^{2i\alpha}$, this becomes

$$4i \int_{|\xi|=1} \frac{z + 1/z}{4 - (z - 1/z)(\xi - 1/\xi)} \frac{d\xi}{\xi},$$

which can be handled by simple fractions.

Alternatively, use the representation below formula (2.9) to reduce the question to the integral which can be evaluated in polar coordinates. Namely, write $\rho = \sin 2\theta$, where $-\pi/2 \le \theta < \pi/2$. Then

$$P(X > 0, Y > 0) = \int_0^{\infty} \int_I \frac{1}{2\pi} r \exp(-r^2/2)\,dr\,d\theta,$$

where $I = \{\alpha \in [-\pi, \pi] : \cos(\alpha - \theta) > 0 \text{ and } \sin(\alpha + \theta) > 0\}$. In particular, for $\theta > 0$ we have $I = (-\theta, \pi/2 + \theta)$ which gives $P(X > 0, Y > 0) = 1/4 + \theta/\pi$.

Problem 2.7 By Corollary 2.3.6 we have $f(t) = \phi(-it) = E\exp(tX) > 0$ for each $t \in \mathbb{R}$, ie. $\log f(t)$ is well defined. By the Cauchy-Schwarz inequality $f(\frac{t+s}{2}) = E\exp(tX/2)\exp(sX/2) \le (f(t)f(s))^{1/2}$, which shows that $\log f(t)$ is convex.
Note: The same is true, but less direct to verify, for the so called analytic ridge functions, see [99].

Problem 2.8 The assumption means that we have independent random variables X_1, X_2 such that $X_1 + X_2 = 1$. Put $Y = X_1 + 1/2, Z = -X_2 - 1/2$. Then Y, Z are independent and $Y = Z$. Hence for any $t \in \mathbb{R}$ we have $P(Y \le t) = P(Y \le t, Z \le t) = P(Y \le t)P(Z \le t) = P(Z \le t)^2$, which is possible only if either $P(Z \le t) = 0$, or $P(Z \le t) = 1$. Since t was arbitrary, the cumulative distribution function of Z has a jump of size 1, i. e. Z is non-random.

For analytic proof, see the solution of Problem 3.6 below.

A.3   Solutions for Chapter 3

Problem 3.1 *Hint:* Show that there is $C > 0$ such that $E\exp(-tX) = C^t$ for all $t \geq 0$. Condition $X \geq 0$ guarantees that Ee^{zX} is analytic for $\Re z < 0$.

Problem 3.4 Write $\mathbf{X} = \mathbf{m} + \mathbf{Y}$. Notice that the characteristic function of $\mathbf{m} - \mathbf{Y}$ and $\mathbf{m} + \mathbf{Y}$ is the same. Therefore $P(\mathbf{m} - \mathbf{Y} \in \mathbb{L}) = P(\mathbf{m} + \mathbf{Y} \in \mathbb{L})$. By Theorem 3.2.1 the probability is either zero (in which case there is nothing to prove) or 1. In the later case, for almost all ω we have $\mathbf{m} + \mathbf{Y} \in \mathbb{L}$ and $\mathbf{m} - \mathbf{Y} \in \mathbb{L}$. But then, the linear combination $\mathbf{m} = \frac{1}{2}(\mathbf{m} + \mathbf{Y}) + \frac{1}{2}(\mathbf{m} - \mathbf{Y}) \in \mathbb{L}$, a contradiction.

Problem 3.5 *Hint:* Show that $Var(X) = 0$.

Problem 3.6 The characteristic functions satisfy $\phi_X(t) = \phi_X(t)\phi_Y(t)$. This shows that $\phi_Y(t) = 1$ in some neighborhood of 0. In particular, $EY^2 = 0$.

For probabilistic proof, see the solution of Problem 2.8.

A.4   Solutions for Chapter 4

Problem 4.1 See, eg. [98].

Problem 4.2 Denote $\rho = corr(X, Y) = \sin\theta$, where $-\pi/2 \leq \theta \leq \pi/2$ By Theorem 4.1.2 we have

$$E\{|\gamma_1| \, |\gamma_1 \cos\theta + \gamma_2 \sin\theta|\}$$

$$= \frac{1}{2\pi} \int_0^{2\pi} |\cos\alpha| \, |\cos\alpha \sin\theta + \sin\alpha \cos\theta| \, d\alpha \int_0^\infty r^3 e^{-r^2/2} \, dr.$$

Therefore $E|X|\,|Y| = \frac{1}{\pi} \int_0^{2\pi} |\cos\alpha||\sin(\alpha + \theta)| \, d\alpha = \frac{1}{2\pi} \int_0^{2\pi} |\sin(2\alpha + \theta) - \sin\theta| \, d\alpha$. Changing the variable of integration to $\beta = 2\alpha$ we have

$$E|X|\,|Y| = \frac{1}{4\pi} \int_0^{4\pi} |\sin(\beta + \theta) - \sin\theta| \, d\beta$$

$$= \frac{1}{2\pi} \int_0^{2\pi} |\sin(\beta + \theta) - \sin\theta \, d\beta.$$

Splitting this into positive and negative parts we get

$$E|X|\,|Y| = \frac{1}{2\pi} \int_0^{\pi - 2\theta} (\sin(\beta + \theta) - \sin\theta) \, d\beta$$

$$-\frac{1}{2\pi} \int_{\pi - 2\theta}^{2\pi} (\sin(\beta + \theta) - \sin\theta) \, d\beta = \frac{2}{\pi}(\cos\theta + \theta\sin\theta).$$

Problem 4.3 *Hint:* Calculate $E|aX + bY|$ in polar coordinates.

Problem 4.5 $E|aX + bY| = 0$ implies $aX + bY = 0$ with probability one. Hence Problem 2.8 implies that both aX and bY are deterministic.

A.5 Solutions for Chapter 5

Problem 5.1 See [111].

Problem 5.2 *Note:* Theorem 6.3.1 gives a stronger result.

Problem 5.3 The joint characteristic function of $X + U, X + V$ is $\phi(t, s) = \psi_X(t+s)\phi_U(t)\phi_V(s)$. On the other hand, by independence of linear forms,

$$\phi(t, s) = \psi_X(t)\phi_X(s)\phi_U(t)\phi_V(s).$$

Therefore for all t, s small enough, we have $\phi_X(t + s) = \phi_X(t)\phi_X(s)$. This shows that there is $\epsilon > 0$ such that $\phi_X(\epsilon 2^{-n}) = C^{2^{-n}}$. Corollary 2.3.4 ends the proof.

Note: This situation is not covered by Theorem 5.3.1 since some of the coefficients in the linear forms are zero.

Problem 5.4 Consider independent random variables $\xi_1 = X - \rho Y, \xi_2 = Y$. Then $X = \xi_1 + \rho\xi_2$ and $Y - \rho X = -\rho\xi_1 + (1 - \rho^2)\xi_2$ are independent linear forms, therefore by Theorem 5.3.1 both ξ_1 and ξ_2 are independent normal random variables. Hence X, Y are jointly normal.

A.6 Solutions for Chapter 6

Problem 6.1 *Hint:* Decompose X, Y into the real and imaginary parts.

Problem 6.2 For standardized one dimensional X, Y with correlation coefficient $\rho \neq 0$ one has $P(X > -M|Y > t) \leq P(X - \rho t > -M)$ which tends to 0 as $t \to \infty$. Therefore $\alpha_{1,0} \geq P(X > -M) - P(X > -M|Y > t)$ has to be 1.

Notice that to prove the result in the general $\mathbb{R}^d \times \mathbb{R}^d$-valued case it is enough to establish stochastic independence of one dimensional variables $\mathbf{u} \cdot \mathbf{X}, \mathbf{v} \cdot \mathbf{Y}$ for all $\mathbf{u}, \mathbf{v} \in \mathbb{R}^d$.

Problem 6.4 Without loss of generality we may assume $Ef(X) = Eg(Y) = 0$. Also, by a linear change of variable if necessary, we may assume $EX = EY = 0$, $EX^2 = EY^2 = 1$. Expanding f, g into Hermite polynomials we have

$$f(x) = \sum_{k=0}^{\infty} f_k/k! H_k(x)$$

$$g(x) = \sum_{k=0}^{\infty} g_k/k! H_k(x)$$

and $\sum f_k^2/k! = Ef(X)^2, \sum g_k^2/k! = Eg(Y)^2$. Moreover, $f_0 = g_0 = 0$ since $Ef(X) = Eg(y) = 0$. Denote by $q(x, y)$ the joint density of X, Y and let $q(\cdot)$ be the marginal density. Mehler's formula (2.12) says that

$$q(x, y) = \sum_{k=0}^{\infty} \rho^k/k! H_k(x)H_k(y)q(x)q(y).$$

Therefore by Cauchy-Schwarz inequality

$$Cov(f, g) = \sum_{k=1}^{\infty} \rho^k/k! f_k g_k \leq |\rho|(\sum f_k^2/k!)^{1/2}(\sum g_k^2/k!)^{1/2}.$$

Problem 6.5 From Problem 6.4 we have $corr(f(X), g(Y)) \leq |\rho|$. Problem 2.3 implies $\frac{|\rho|}{2\pi} \leq \frac{1}{2\pi} \arcsin|\rho| \leq P(X > 0, \pm Y > 0) - P(X > 0)P(\pm Y > 0) \leq \alpha_{0,0}$.

For the general case see, eg. [128, page 74 Lemma 2].

Problem 6.6 *Hint:* Follow the proof of Theorem 6.2.2. A slightly more general proof can be found in [19, Theorem A].

Problem 6.7 *Hint:* Use the tail integration formula (1.2) and estimate (6.8), see Problem 1.5.

A.7   Solutions for Chapter 7

Problem 7.1 Since $(X_1, X_1+X_2) \cong (X_2, X_1+X_2)$, we have $E\{X_1|X_1+X_2\} = E\{X_2|X_1+X_2\}$, cf. Problem 1.11.

Problem 7.2 By symmetry of distributions, $(X_1 + X_2, X_1 - X_2) \cong (X_1 - X_2, X_1 + X_2)$. Therefore $E\{X_1+X_2|X_1-X_2\} = 0$, see Problem 7.1 and the result follows from Theorem 7.1.2.

Problem 7.5 (i) If Y is degenerated, then $a = 0$, see Problem 7.3. For non-degenerated Y the conclusion follows from Problem 1.12, since by independence $E\{X + Y|Y\} = X$.
(ii) Clearly, $E\{X|X + Y\} = (1 - a)(X + Y)$. Therefore $Y = \frac{1}{1-a}E\{X|X + Y\} - X$ and $\|Y\|_p \leq \frac{2}{1-a}\|X\|_p$.

Problem 7.6 Problem 7.5 implies that Y has all moments and a can be expressed explicitly by the variances of X, Y. Let Z be independent of X normal such that $E\{Z|Z + X\} = a(X + Z)$ with the same a (ie. $Var(Z) = Var(Y)$). Since the normal distribution is uniquely determined by moments, it is enough to show that all moments of Y are uniquely determined (as then they have to equal to the corresponding moments of Z).

To this end write $EY(X + Y)^n = aE(X + Y)^{n+1}$, which gives $(1 - a)EY^{n+1} = \sum_{k=0}^{n} a \binom{n+1}{k} EY^k EX^{n+1-k} - \sum_{k=0}^{n-1} \binom{n}{k} EY^{k+1} EX^{n-k}$.

Problem 7.7 It is obvious that $E\{X|Y\} = \rho Y$, because Y has two values only, and two points are always on some straight line; alternatively write the joint characteristic function.

Formula $Var(X|Y) = 1 - \rho^2$ follows from the fact that the conditional distribution of X given $Y = 1$ is the same as the conditional distribution of $-X$ given $Y = -1$; alternatively, write the joint characteristic function and use Theorem 1.5.3. The other two relations follow from $(X, Y) \cong (Y, X)$.

Problem 7.3 Without loss of generality we may assume $EX = 0$. Put $U = Y, V = X + Y$. By independence, $E\{V|U\} = U$. On the other hand $E\{U|V\} = E\{X + Y - X|X + Y\} = X + Y = V$. Therefore by Problem 1.14, $X + Y \cong Y$ and $X = 0$ by Problem 3.6.

Problem 7.4 Without loss of generality we may assume $EX = 0$. Then $EU = 0$. By Jensen's inequality $EX^2 + EY^2 = E(X + Y)^2 \leq EX^2$, so $EY^2 = 0$.

Problem 7.8 This follows the proof of Theorem 7.1.2 and Lemma 7.3.2. Explicit computation is in [21, Lemma 2.1].

Problem 7.9 This follows the proof of Theorem 7.1.2 and Lemma 7.3.2. Explicit computation is in [148, Lemma 2.3].

Bibliography

[1] J. Aczél, *Lectures on functional equations and their applications*, Academic Press, New York 1966.

[2] N. I. Akhiezer, *The Classical Moment Problem*, Oliver & Boyd, Edinburgh, 1965.

[3] C. D. Aliprantis & O. Burkinshaw, *Positive operators*, Acad. Press 1985.

[4] T. A. Azlarov & N. A. Volodin, *Characterization Problems Associated with the Exponential Distribution*, Springer, New York 1986.

[5] N. Aronszajn, Theory of reproducing kernels, *Trans. Amer. Math. Soc.* 68 (1950), pp. 337–404.

[6] M. S. Barlett, The vector representation of the sample, *Math. Proc. Cambr. Phil. Soc.* 30 (1934), pp. 327–340.

[7] A. Bensoussan, *Stochastic Control by Functional Analysis Methods*, North-Holland, Amsterdam 1982.

[8] S. N. Bernstein, On the property characteristic of the normal law, *Trudy Leningrad. Polytechn. Inst.* 3 (1941), pp. 21–22; *Collected papers*, Vol IV, Nauka, Moscow 1964, pp. 314–315.

[9] P. Billingsley, *Probability and measure*, Wiley, New York 1986.

[10] P. Billingsley, *Convergence of probability measures*, Wiley New York 1968.

[11] S. Bochner, Über eine Klasse singulärer Integralgleichungen, *Berliner Sitzungsberichte* 22 (1930), pp. 403–411.

[12] E. M. Bolger & W. L. Harkness, Characterizations of some distributions by conditional moments, *Ann. Math. Statist.* 36 (1965), pp. 703–705.

[13] R. C. Bradley & W. Bryc, Multilinear forms and measures of dependence between random variables *Journ. Multiv. Analysis* 16 (1985), pp. 335–367.

[14] M. S. Braverman, Characteristic properties of normal and stable distributions, *Theor. Probab. Appl.* 30 (1985), pp. 465–474.

[15] M. S. Braverman, On a method of characterization of probability distributions, *Theor. Probab. Appl.* 32 (1987), pp. 552–556.

[16] M. S. Braverman, Remarks on Characterization of Normal and Stable Distributions, *Journ. Theor. Probab.* 6 (1993), pp. 407–415.

[17] A. Brunnschweiler, A connection between the Boltzman Equation and the Ornstein-Uhlenbeck Process, *Arch. Rational Mechanics* 76 (1980), pp. 247–263.

[18] W. Bryc & P. Szablowski, Some characteristic of normal distribution by conditional moments I, *Bull. Polish Acad. Sci. Ser. Math.* 38 (1990), pp. 209–218.

[19] W. Bryc, Some remarks on random vectors with nice enough behaviour of conditional moments, *Bull. Polish Acad. Sci. Ser. Math.* 33 (1985), pp. 677–683.

[20] W. Bryc & A. Plucińska, A characterization of infinite Gaussian sequences by conditional moments, *Sankhya, Ser. A* 47 (1985), pp. 166–173.

[21] W. Bryc, A characterization of the Poisson process by conditional moments, *Stochastics* 20 (1987), pp. 17–26.

[22] W. Bryc, Remarks on properties of probability distributions determined by conditional moments, *Probab. Theor. Rel. Fields* 78 (1988), pp. 51–62.

[23] W. Bryc, On bivariate distributions with "rotation invariant" absolute moments, *Sankhya, Ser. A* 54 (1992), pp. 432–439.

[24] T. Byczkowski & T. Inglot, Gaussian Random Series on Metric Vector Spaces, *Mathematische Zeitschrift* 196 (1987), pp. 39–50.

[25] S. Cambanis, S. Huang & G. Simons, On the theory of elliptically contoured distributions, *Journ. Multiv. Analysis* 11 (1981), pp. 368–385.

[26] R. V. Chacon & J. B. Walsh, One dimensional potential embedding, Seminaire de Probabilities X, *Lecture Notes in Math.* 511 (1976), pp. 19–23.

[27] L. Corwin, Generalized Gaussian measures and a functional equation, *J. Functional Anal.* 5 (1970), pp. 412–427.

[28] H. Cramer, On a new limit theorem in the theory of probability, in *Colloquium on the Theory of Probability*, Hermann, Paris 1937.

[29] H. Cramer, Uber eine Eigenschaft der normalen Verteilungsfunktion, *Math. Zeitschrift* 41 (1936), pp. 405–411.

[30] G. Darmois, Analyse générale des liasons stochastiques, *Rev. Inst. Internationale Statist.* 21 (1953), pp. 2–8.

[31] B. de Finetti, Funzione caratteristica di un fenomeno aleatorio, *Memorie Rend. Accad. Lincei* 4 (1930), pp. 86–133.

[32] A. Dembo & O. Zeitouni, *Large Deviation Techniques and Applications*, Jones and Bartlett Publ., Boston 1993.

[33] J. Deny, Sur l'equation de convolution $\mu = \mu \star \sigma$, *Seminaire BRELOT-CHOCQUET-DENY (theorie du Potentiel)* 4e anneé, 1959/60.

[34] J-D. Deuschel & D. W. Stroock, *Introduction to Large Deviations*, Acad. Press, Boston 1989.

[35] P. Diaconis & D. Freedman, A dozen of de Finetti-style results in search of a theory, *Ann. Inst. Poincaré, Probab. & Statist.* 23 (1987), pp. 397–422.

[36] P. Diaconis & D. Ylvisaker, Quantifying prior opinion, *Bayesian Statistics 2*, Eslevier Sci. Publ. 1985 (Editors: J. M. Bernardo et al.), pp. 133–156.

[37] P. L. Dobruschin, The description of a random field by means of conditional probabilities and conditions of its regularity, *Theor. Probab. Appl.* 13 (1968), pp. 197–224.

[38] J. Doob, *Stochastic processes*, Wiley, New York 1953.

[39] P. Doukhan, *Mixing: properties and examples*, Lecture Notes in Statistics, Springer, Vol. 85 (1994).

[40] M. Dozzi, *Stochastic processes with multidimensional parameter*, Pitman Research Notes in Math., Longman, Harlow 1989.

[41] R. Durrett, *Brownian Motion and Martingales in Analysis*, Wadsworth, Belmont, Ca 1984.

[42] R. Durrett, *Probability: Theory and examples*, Wadsworth, Belmont, Ca 1991.

[43] N. Dunford & J. T. Schwartz, *Linear Operators I*, Interscience, New York 1958.

[44] O. Enchev, Pathwise nonlinear filtering on Abstract Wiener Spaces, *Ann. Probab.* 21 (1993), pp. 1728–1754.

[45] C. G. Esseen, Fourier analysis of distribution functions, *Acta Math.* 77 (1945), pp. 1–124.

[46] S. N. Ethier & T. G. Kurtz, *Markov Processes*, Wiley, New York 1986.

[47] K. T. Fang & T. W. Anderson, *Statistical inference in elliptically contoured and related distributions*, Allerton Press, Inc., New York 1990.

[48] K. T. Fang, S. Kotz & K.-W. Ng, *Symmetric Multivariate and Related Distributions*, Monographs on Statistics and Applied Probability 36, Chapman and Hall, London 1990.

[49] G. M. Feldman, *Arithmetics of probability distributions and characterization problems*, Naukova Dumka, Kiev 1990 (in Russian).

[50] G. M. Feldman, Marcinkiewicz and Lukacs Theorems on Abelian Groups, *Theor. Probab. Appl.* 34 (1989), pp. 290–297.

[51] G. M. Feldman, Bernstein Gaussian distributions on groups, *Theor. Probab. Appl.*
31 (1986), pp. 40–49; *Teor. Verojatn. Primen.* 31 (1986), pp. 47–58.

[52] G. M. Feldman, Characterization of the Gaussian distribution on groups by independence of linear statistics, *DAN SSSR* 301 (1988), pp. 558–560; English translation *Soviet-Math.-Doklady* 38 (1989), pp. 131–133.

[53] W. Feller, On the logistic law of growth and its empirical verification in biology, *Acta Biotheoretica* 5 (1940), pp. 51–55.

[54] W. Feller, *An Introduction to Probability Theory*, Vol. II, Wiley, New York 1966.

[55] X. Fernique, *Régularité des trajectories des fonctions aléatoires gaussienes*, In: *Ecoles d'Eté de Probabilités de Saint-Flour IV – 1974, Lecture Notes in Math.*, Springer, Vol. 480 (1975), pp. 1–97.

[56] X. Fernique, Integrabilite des vecteurs gaussienes, *C. R. Acad. Sci. Paris*, Ser. A, 270 (1970), pp. 1698–1699.

[57] J. Galambos & S. Kotz, *Characterizations of Probability Distributions, Lecture Notes in Math.*, Springer, Vol. 675 (1978).

[58] P. Hall, A distribution is completely determined by its translated moments, *Zeitschr. Wahrscheinlichkeitsth. v. Geb.* 62 (1983), pp. 355–359.

[59] P. R. Halmos, *Measure Theory*, Van Nostrand Reinhold Co, New York 1950.

[60] C. Hardin, Isometries on subspaces of L_p, *Indiana Univ. Math. Journ.* 30 (1981), pp. 449–465.

[61] C. Hardin, On the linearity of regression, *Zeitschr. Wahrscheinlichkeitsth. v. Geb.* 61 (1982), pp. 293–302.

[62] G. H. Hardy & E. C. Titchmarsh, An integral equation, *Math. Proc. Cambridge Phil. Soc.* 28 (1932), pp. 165–173.

[63] J. F. W. Herschel, Quetelet on Probabilities, *Edinburgh Rev.* 92 (1850) pp. 1–57.

[64] W. Hoeffding, Masstabinvariante Korrelations-theorie, *Schriften Math. Inst. Univ. Berlin* 5 (1940) pp. 181–233.

[65] I. A. Ibragimov & Yu. V. Linnik, *Independent and stationary sequences of random variables*, Wolters-Nordhoff, Gröningen 1971.

[66] K. Ito & H. McKean *Diffusion processes and their sample paths*, Springer, New York 1964.

[67] J. Jakubowski & S. Kwapień, On multiplicative systems of functions, *Bull. Acad. Polon. Sci. (Bull. Polish Acad. Sci.)*, Ser. Math. 27 (1979) pp. 689–694.

[68] S. Janson, Normal convergence by higher semiinvariants with applications to sums of dependent random variables and random graphs, *Ann. Probab.* 16 (1988), pp. 305–312.

[69] N. L. Johnson & S. Kotz, *Distributions in Statistics: Continuous Multivariate Distributions*, Wiley, New York 1972.

[70] N. L. Johnson & S. Kotz & A. W. Kemp, *Univariate discrete distributions*, Wiley, New York 1992.

[71] M. Kac, On a characterization of the normal distribution, *Amer. J. Math.* 61 (1939), pp. 726–728.

[72] M. Kac, O stochastycznej niezależnosci funkcyj, *Wiadomości Matematyczne* 44 (1938), pp. 83–112 (in Polish).

[73] A. M. Kagan, Ju. V. Linnik & C. R. Rao, *Characterization Problems of Mathematical Statistics*, Wiley, New York 1973.

[74] A. M. Kagan, A Generalized Condition for Random Vectors to be Identically Distributed Related to the Analytical Theory of Linear Forms of Independent Random Variables, *Theor. Probab. Appl.* 34 (1989), pp. 327–331.

[75] A. M. Kagan, The Lukacs-King method applied to problems involving linear forms of independent random variables, *Metron*, Vol. XLVI (1988), pp. 5–19.

[76] J. P. Kahane, *Some random series of functions*, Cambridge University Press, New York 1985.

[77] J. F. C. Kingman, Random variables with unsymmetrical linear regressions, *Math. Proc. Cambridge Phil. Soc.*, 98 (1985), pp. 355–365.

[78] J. F. C. Kingman, The construction of infinite collection of random variables with linear regressions, *Adv. Appl. Probab.* 1986 Suppl., pp. 73–85.

[79] *Exchangeability in probability and statistics*, edited by G. Koch & F. Spizzichino, North-Holland, Amsterdam 1982.

[80] A. L. Koldobskii, Convolution equations in certain Banach spaces, *Proc. Amer. Math. Soc.* 111 (1991), pp. 755–765.

[81] A. L. Koldobskii, The Fourier transform technique for convolution equations in infinite dimensional ℓ_q-spaces, *Mathematische Annalen* 291 (1991), pp. 403–407.

[82] A. N. Kolmogorov, *Foundations of the Theory of Probability*, Chelsea, New York 1956.

[83] A. N. Kolmogorov & Yu. A. Rozanov, On strong mixing conditions for stationary Gaussian processes, *Theory Probab. Appl.* 5 (1960), pp. 204–208.

[84] I. I. Kotlarski, On characterizing the gamma and the normal distribution, *Pacific Journ. Math.* 20 (1967), pp. 69–76.

[85] I. I. Kotlarski, On characterizing the normal distribution by Student's law, *Biometrika* 53 (1963), pp. 603–606.

[86] I. I. Kotlarski, On a characterization of probability distributions by the joint distribution of their linear functions, *Sankhya* Ser. A, 33 (1971), pp. 73–80.

[87] I. I. Kotlarski, Explicit formulas for characterizations of probability distributions by using maxima of a random number of random variables, *Sankhya* Ser. A Vol. 47 (1985), pp. 406–409.

[88] W. Krakowiak, Zero-one laws for A-decomposable measures on Banach spaces, *Bull. Polish Acad. Sci.* Ser. Math. 33 (1985), pp. 85–90.

[89] W. Krakowiak, The theorem of Darmois-Skitovich for Banach space valued random variables, *Bull. Polish Acad. Sci.* Ser. Math. 33 (1985), pp. 77–85.

[90] H. H. Kuo, *Gaussian Measures in Banach Spaces, Lecture Notes in Math.*, Springer, Vol. 463 (1975).

[91] S. Kwapień, Decoupling inequalities for polynomial chaos, *Ann. Probab.* 15 (1987), pp. 1062–1071.

[92] R. G. Laha, On the laws of Cauchy and Gauss, *Ann. Math. Statist.* 30 (1959), pp. 1165–1174.

[93] R. G. Laha, On a characterization of the normal distribution from properties of suitable linear statistics, *Ann. Math. Statist.* 28 (1957), pp. 126–139.

[94] H. O. Lancaster, Joint Probability Distributions in the Meixner Classes, *J. Roy. Statist. Assoc.* Ser. B 37 (1975), pp. 434–443.

[95] V. P. Leonov & A. N. Shiryaev, Some problems in the spectral theory of higher-order moments, II, *Theor. Probab. Appl.* 5 (1959), pp. 417–421.

[96] G. Letac, Isotropy and sphericity: some characterizations of the normal distribution, *Ann. Statist.* 9 (1981), pp. 408–417.

[97] B. Ja. Levin, *Distribution of zeros of entire functions*, Transl. Math. Monographs Vol. 5, Amer. Math. Soc., Providence, Amer. Math. Soc. 1972.

[98] P. Levy, *Théorie de l'addition des variables aléatoires*, Gauthier-Villars, Paris 1937.

[99] Ju. V. Linnik & I. V. Ostrovskii, *Decompositions of random variables and vectors*, Providence, Amer. Math. Soc. 1977.

[100] Ju. V. Linnik, O nekatorih odinakovo raspredelenyh statistikah, *DAN SSSR (Sovet Mathematics - Doklady)* 89 (1953), pp. 9–11.

[101] E. Lukacs & R. G. Laha, *Applications of Characteristic Functions*, Hafner Publ, New York 1964.

[102] E. Lukacs, Stability Theorems, *Adv. Appl. Probab.* 9 (1977), pp. 336–361.

[103] E. Lukacs, *Characteristic Functions*, Griffin & Co., London 1960.

[104] E. Lukacs & E. P. King, A property of the normal distribution, *Ann. Math. Statist.* 13 (1954), pp. 389–394.

[105] J. Marcinkiewicz, *Collected Papers*, PWN, Warsaw 1964.

[106] J. Marcinkiewicz, Sur une propriete de la loi de Gauss, *Math. Zeitschrift* 44 (1938), pp. 612–618.

[107] J. Marcinkiewicz, Sur les variables aléatoires enroulées, *C. R. Soc. Math. France* Anneé 1938, (1939), pp. 34–36.

[108] J. C. Maxwell, *Illustrations of the Dynamical Theory of Gases*, Phil. Mag. 19 (1860), pp. 19–32. Reprinted in *The Scientific Papers of James Clerk Maxwell*, Vol. I, Edited by W. D. Niven, Cambridge, University Press 1890, pp. 377–409.

[109] *Maxwell on Molecules and Gases*, Edited by E. Garber, S. G. Brush & C. W. Everitt, 1986, Massachusetts Institute of Technology.

[110] W. Magnus & F. Oberhettinger, *Formulas and theorems for the special functions of mathematical physics*, Chelsea, New York 1949.

[111] V. Menon & V. Seshadri, The Darmois-Skitovic theorem characterizing the normal law, *Sankhya* 47 (1985) Ser. A, pp. 291–294.

[112] L. D. Meshalkin, On the robustness of some characterizations of the normal distribution, *Ann. Math. Statist.* 39 (1968), pp. 1747–1750.

[113] K. S. Miller, *Multidimensional Gaussian Distributions*, Wiley, New York 1964.

[114] S. Narumi, On the general forms of bivariate frequency distributions which are mathematically possible when regression and variation are subject to limiting conditions I, II, *Biometrika* 15 (1923), pp. 77–88; 209–221.

[115] E. Nelson, *Dynamical theories of Brownian motion*, Princeton Univ. Press 1967.

[116] E. Nelson, The free Markov field, *J. Functional Anal.* 12 (1973), pp. 211–227.

[117] J. Neveu, *Discrete-parameter martingales*, North-Holland, Amsterdam 1975.

[118] I. Nimo-Smith, Linear regressions and sphericity, *Biometrica* 66 (1979), pp. 390–392.

[119] R. P. Pakshirajan & N. R. Mohun, A characterization of the normal law, *Ann. Inst. Statist. Math.* 21 (1969), pp. 529–532.

[120] J. K. Patel & C. B. Read, *Handbook of the normal distribution*, Dekker, New York 1982.

[121] A. Plucińska, On a stochastic process determined by the conditional expectation and the conditional variance, *Stochastics* 10 (1983), pp. 115–129.

[122] G. Polya, Herleitung des Gauss'schen Fehlergesetzes aus einer Funktionalgleichung, *Math. Zeitschrift* 18 (1923), pp. 96–108.

[123] S. C. Port & C. J. Stone, *Brownian motion and classical potential theory* Acad. Press, New York 1978.

[124] Yu. V. Prokhorov, Characterization of a class of distributions through the distribution of certain statistics, *Theor. Probab. Appl.* 10 (1965), pp. 438-445; *Teor. Ver.* 10 (1965), pp. 479–487.

[125] B. L. S. Prakasa Rao, *Identifiability in Stochastic Models* Acad. Press, Boston 1992.

[126] B. Ramachandran & B. L. S. Praksa Rao, On the equation $f(x) = \int_{-\infty}^{\infty} f(x+y)\mu(dy)$, *Sankhya* Ser. A, 46 (1984,) pp. 326–338.

[127] D. Richardson, Random growth in a tessellation, *Math. Proc. Cambridge Phil. Soc.* 74 (1973), pp. 515–528.

[128] M. Rosenblatt, *Stationary sequences and random fields*, Birkhäuser, Boston 1985.

[129] W. Rudin, L_p-isometries and equimeasurability, *Indiana Univ. Math. Journ.* 25 (1976), pp. 215–228.

[130] R. Sikorski, *Advanced Calculus*, PWN, Warsaw 1969.

[131] D. N. Shanbhag, An extension of Lukacs's result, *Math. Proc. Cambridge Phil. Soc.* 69 (1971), pp. 301–303.

[132] I. J. Schönberg, Metric spaces and completely monotone functions, *Ann. Math.* 39 (1938), pp. 811–841.

[133] R. Shimizu, Characteristic functions satisfying a functional equation I, *Ann. Inst. Statist. Math.* 20 (1968), pp. 187–209.

[134] R. Shimizu, Characteristic functions satisfying a functional equation II, *Ann. Inst. Statist. Math.* 21 (1969), pp. 395–405.

[135] R. Shimizu, Correction to: "Characteristic functions satisfying a functional equation II", *Ann. Inst. Statist. Math.* 22 (1970), pp. 185–186.

[136] V. P. Skitovich, On a property of the normal distribution, *DAN SSSR* (Doklady) 89 (1953), pp. 205–218.

[137] A. V. Skorohod, *Studies in the theory of random processes*, Addison-Wesley, Reading, Massachusetts 1965. (Russian edition: 1961.)

[138] W. Smoleński, A new proof of the zero-one law for stable measures, *Proc. Amer. Math. Soc.* 83 (1981), pp. 398–399.

[139] P. J. Szabłowski, Can the first two conditional moments identify a mean square differentiable process?, *Computers Math. Applic.* 18 (1989), pp. 329–348.

[140] P. Szabłowski, Some remarks on two dimensional elliptically contoured measures with second moments, *Demonstr. Math.* 19 (1986), pp. 915–929.

[141] P. Szabłowski, Expansions of $E\{X|Y + \epsilon Z\}$ and their applications to the analysis of elliptically contoured measures, *Computers Math. Applic.* 19 (1990), pp. 75–83.

[142] P. J. Szabłowski, *Conditional Moments and the Problems of Characterization*, preprint 1992.

[143] A. Tortrat, Lois stables dans un groupe, *Ann. Inst. H. Poincaré Probab. & Statist.* 17 (1981), pp. 51–61.

[144] E. C. Titchmarsh, *The theory of functions*, Oxford Univ. Press, London 1939.

[145] Y. L. Tong, *The Multivariate Normal Distribution*, Springer, New York 1990.

[146] K. Urbanik, Random linear functionals and random integrals, *Colloquium Mathematicum* 33 (1975), pp. 255–263.

[147] J. Wesołowski, Characterizations of some Processes by properties of Conditional Moments, *Demonstr. Math.* 22 (1989), pp. 537–556.

[148] J. Wesołowski, A Characterization of the Gamma Process by Conditional Moments, *Metrika* 36 (1989) pp. 299–309.

[149] J. Wesołowski, Stochastic Processes with linear conditional expectation and quadratic conditional variance, *Probab. Math. Statist. (Wrocław)* 14 (1993) pp. 33–44.

[150] N. Wiener, Differential space, *J. Math. Phys. MIT* 2 (1923), pp. 131–174.

[151] Z. Zasvari, Characterizing distributions of the random variables X_1, X_2, X_3 by the distribution of $(X_1 - X_2, X_2 - X_3)$, *Probab. Theory Rel. Fields* 73 (1986), pp. 43–49.

[152] V. M. Zolotariev, Mellin-Stjelties transforms in probability theory, *Theor. Probab. Appl.* 2 (1957), pp. 433–460.

[153] Proceedings of 7-th Oberwolfach conference on probability measures on groups (1981), *Lecture Notes in Math.*, Springer, Vol. 928 (1982).

Index

Lecture Notes in Statistics

For information about Volumes 1 to 12
please contact Springer-Verlag

Vol. 13: J. Pfanzagl (with the assistance of W. Wefelmeyer), Contributions to a General Asymptotic Statistical Theory. vii, 315 pages, 1982.

Vol. 14: GLIM 82: Proceedings of the International Conference on Generalised Linear Models. Edited by R. Gilchrist. v, 188 pages, 1982.

Vol. 15: K.R.W. Brewer and M. Hanif, Sampling with Unequal Probabilities. ix, 164 pages, 1983.

Vol. 16: Specifying Statistical Models: From Parametric to Non-Parametric, Using Bayesian or Non-Bayesian Approaches. Edited by J.P. Florens, M. Mouchart, J.P. Raoult, L. Simar, and A.F.M. Smith, xi, 204 pages, 1983.

Vol. 17: I.V. Basawa and D.J. Scott, Asymptotic Optimal Inference for Non-Ergodic Models. ix, 170 pages, 1983.

Vol. 18: W. Britton, Conjugate Duality and the Exponential Fourier Spectrum. v, 226 pages, 1983.

Vol. 19: L. Fernholz, von Mises Calculus For Statistical Functionals. viii, 124 pages, 1983.

Vol. 20: Mathematical Learning Models — Theory and Algorithms: Proceedings of a Conference. Edited by U. Herkenrath, D. Kalin, W. Vogel. xiv, 226 pages, 1983.

Vol. 21: H. Tong, Threshold Models in Non-linear Time Series Analysis. x, 323 pages, 1983.

Vol. 22: S. Johansen, Functional Relations, Random Coefficients and Nonlinear Regression with Application to Kinetic Data, viii, 126 pages, 1984.

Vol. 23: D.G. Saphire, Estimation of Victimization Prevalence Using Data from the National Crime Survey. v, 165 pages, 1984.

Vol. 24: T.S. Rao, M.M. Gabr, An Introduction to Bispectral Analysis and Bilinear Time Series Models. viii, 280 pages, 1984.

Vol. 25: Time Series Analysis of Irregularly Observed Data. Proceedings, 1983. Edited by E. Parzen. vii, 363 pages, 1984.

Vol. 26: Robust and Nonlinear Time Series Analysis. Proceedings, 1983. Edited by J. Franke, W. Härdle and D. Martin. ix, 286 pages, 1984.

Vol. 27: A. Janssen, H. Milbrodt, H. Strasser, Infinitely Divisible Statistical Experiments. vi, 163 pages, 1985.

Vol. 28: S. Amari, Differential-Geometrical Methods in Statistics. v, 290 pages, 1985.

Vol. 29: Statistics in Ornithology. Edited by B.J.T. Morgan and P.M. North. xxv, 418 pages, 1985.

Vol 30: J. Grandell, Stochastic Models of Air Pollutant Concentration. v, 110 pages, 1985.

Vol. 31: J. Pfanzagl, Asymptotic Expansions for General Statistical Models. vii, 505 pages, 1985.

Vol. 32: Generalized Linear Models. Proceedings, 1985. Edited by R. Gilchrist, B. Francis and J. Whittaker. vi, 178 pages, 1985.

Vol. 33: M. Csörgo, S. Csörgo, L. Horváth, An Asymptotic Theory for Empirical Reliability and Concentration Processes. v, 171 pages, 1986.

Vol. 34: D.E. Critchlow, Metric Methods for Analyzing Partially Ranked Data. x, 216 pages, 1985.

Vol. 35: Linear Statistical Inference. Proceedings, 1984. Edited by T. Calinski and W. Klonecki. vi, 318 pages, 1985.

Vol. 36: B. Matérn, Spatial Variation. Second Edition. 151 pages, 1986.

Vol. 37: Advances in Order Restricted Statistical Inference. Proceedings, 1985. Edited by R. Dykstra, T. Robertson and F.T. Wright. viii, 295 pages, 1986.

Vol. 38: Survey Research Designs: Towards a Better Understanding of Their Costs and Benefits. Edited by R.W. Pearson and R.F. Boruch. v, 129 pages, 1986.

Vol. 39: J.D. Malley, Optimal Unbiased Estimation of Variance Components. ix, 146 pages, 1986.

Vol. 40: H.R. Lerche, Boundary Crossing of Brownian Motion. v, 142 pages, 1986.

Vol. 41: F. Baccelli, P. Brémaud, Palm Probabilities and Stationary Queues. vii, 106 pages, 1987.

Vol. 42: S. Kullback, J.C. Keegel, J.H. Kullback, Topics in Statistical Information Theory. ix, 158 pages, 1987.

Vol. 43: B.C. Arnold, Majorization and the Lorenz Order: A Brief Introduction. vi, 122 pages, 1987.

Vol. 44: D.L. McLeish, Christopher G. Small, The Theory and Applications of Statistical Inference Functions. vi, 124 pages, 1987.

Vol. 45: J.K. Ghosh (Editor), Statistical Information and Likelihood. 384 pages, 1988.

Vol. 46: H.-G. Müller, Nonparametric Regression Analysis of Longitudinal Data. vi, 199 pages, 1988.

Vol. 47: A.J. Getson, F.C. Hsuan, {2}-Inverses and Their Statistical Application. viii, 110 pages, 1988.

Vol. 48: G.L. Bretthorst, Bayesian Spectrum Analysis and Parameter Estimation. xii, 209 pages, 1988.

Vol. 49: S.L. Lauritzen, Extremal Families and Systems of Sufficient Statistics. xv, 268 pages, 1988.

Vol. 50: O.E. Barndorff-Nielsen, Parametric Statistical Models and Likelihood. vii, 276 pages, 1988.

Vol. 51: J. Hüsler, R.-D. Reiss (Editors), Extreme Value Theory. Proceedings, 1987. x, 279 pages, 1989.

Vol. 52: P.K. Goel, T. Ramalingam, The Matching Methodology: Some Statistical Properties. viii, 152 pages, 1989.

Vol. 53: B.C. Arnold, N. Balakrishnan, Relations, Bounds and Approximations for Order Statistics. ix, 173 pages, 1989.

Vol. 54: K.R. Shah, B.K. Sinha, Theory of Optimal Designs. viii, 171 pages, 1989.

Vol. 55: L. McDonald, B. Manly, J. Lockwood, J. Logan (Editors), Estimation and Analysis of Insect Populations. Proceedings, 1988. xiv, 492 pages, 1989.

Vol. 56: J.K. Lindsey, The Analysis of Categorical Data Using GLIM. v, 168 pages, 1989.

Vol. 57: A. Decarli, B.J. Francis, R. Gilchrist, G.U.H. Seeber (Editors), Statistical Modelling. Proceedings, 1989. ix, 343 pages, 1989.

Vol. 58: O.E. Barndorff-Nielsen, P. Blæsild, P.S. Eriksen, Decomposition and Invariance of Measures, and Statistical Transformation Models. v, 147 pages, 1989.

Vol. 59: S. Gupta, R. Mukerjee, A Calculus for Factorial Arrangements. vi, 126 pages, 1989.

Vol. 60: L. Györfi, W. Härdle, P. Sarda, Ph. Vieu, Nonparametric Curve Estimation from Time Series. viii, 153 pages, 1989.

Vol. 61: J. Breckling, The Analysis of Directional Time Series: Applications to Wind Speed and Direction. viii, 238 pages, 1989.

Vol. 62: J.C. Akkerboom, Testing Problems with Linear or Angular Inequality Constraints. xii, 291 pages, 1990.

Vol. 63: J. Pfanzagl, Estimation in Semiparametric Models: Some Recent Developments. iv, 112 pages, 1990.

Vol. 64: S. Gabler, Minimax Solutions in Sampling from Finite Populations. v, 132 pages, 1990.

Vol. 65: A. Janssen, D.M. Mason, Non-Standard Rank Tests. vi, 252 pages, 1990.

Vol. 66: T. Wright, Exact Confidence Bounds when Sampling from Small Finite Universes. xvi, 431 pages, 1991.

Vol. 67: M.A. Tanner, Tools for Statistical Inference: Observed Data and Data Augmentation Methods. vi, 110 pages, 1991.

Vol. 68: M. Taniguchi, Higher Order Asymptotic Theory for Time Series Analysis. viii, 160 pages, 1991.

Vol. 69: N.J.D. Nagelkerke, Maximum Likelihood Estimation of Functional Relationships. v, 110 pages, 1992.

Vol. 70: K. Iida, Studies on the Optimal Search Plan. viii, 130 pages, 1992.

Vol. 71: E.M.R.A. Engel, A Road to Randomness in Physical Systems. ix, 155 pages, 1992.

Vol. 72: J.K. Lindsey, The Analysis of Stochastic Processes using GLIM. vi, 294 pages, 1992.

Vol. 73: B.C. Arnold, E. Castillo, J.-M. Sarabia, Conditionally Specified Distributions. xiii, 151 pages, 1992.

Vol. 74: P. Barone, A. Frigessi, M. Piccioni (Editors), Stochastic Models, Statistical Methods, and Algorithms in Image Analysis. vi, 258 pages, 1992.

Vol. 75: P.K. Goel, N.S. Iyengar (Editors), Bayesian Analysis in Statistics and Econometrics. xi, 410 pages, 1992.

Vol. 76: L. Bondesson, Generalized Gamma Convolutions and Related Classes of Distributions and Densities. viii, 173 pages, 1992.

Vol. 77: E. Mammen, When Does Bootstrap Work? Asymptotic Results and Simulations. vi, 196 pages, 1992.

Vol. 78: L. Fahrmeir, B. Francis, R. Gilchrist, G. Tutz (Editors), Advances in GLIM and Statistical Modelling: Proceedings of the GLIM92 Conference and the 7th International Workshop on Statistical Modelling, Munich, 13-17 July 1992. ix, 225 pages, 1992.

Vol. 79: N. Schmitz, Optimal Sequentially Planned Decision Procedures. xii, 209 pages, 1992.

Vol. 80: M. Fligner, J. Verducci (Editors), Probability Models and Statistical Analyses for Ranking Data. xxii, 306 pages, 1992.

Vol. 81: P. Spirtes, C. Glymour, R. Scheines, Causation, Prediction, and Search. xxiii, 526 pages, 1993.

Vol. 82: A. Korostelev and A. Tsybakov, Minimax Theory of Image Reconstruction. xii, 268 pages, 1993.

Vol. 83: C. Gatsonis, J. Hodges, R. Kass, N. Singpurwalla (Editors), Case Studies in Bayesian Statistics. xii, 437 pages, 1993.

Vol. 84: S. Yamada, Pivotal Measures in Statistical Experiments and Sufficiency. vii, 129 pages, 1994.

Vol. 85: P. Doukhan, Mixing: Properties and Examples. xi, 142 pages, 1994.

Vol. 86: W. Vach, Logistic Regression with Missing Values in the Covariates. xi, 139 pages, 1994.

Vol. 87: J. Møller, Lectures on Random Voronoi Tessellations. vii, 134 pages, 1994.

Vol. 88: J.E. Kolassa, Series Approximation Methods in Statistics. viii, 150 pages, 1994.

Vol. 89: P. Cheeseman and R.W. Oldford (Editors), Selecting Models From Data: Artificial Intelligence and Statistics IV. x, 487 pages, 1994.

Vol. 90: A. Csenki, Dependability for Systems with a Partitioned State Space: Markov and Semi-Markov Theory and Computational Implementation. x, 241 pages, 1994.

Vol. 91: J.D. Malley, Statistical Applications of Jordan Algebras. viii, 101 pages, 1994.

Vol. 92: M. Eerola, Probabilistic Causality in Longitudinal Studies. vii, 133 pages, 1994.

Vol. 93: Bernard Van Cutsem (Editor), Classification and Dissimilarity Analysis. xiv, 238 pages, 1994.

Vol. 94: Jane F. Gentleman and G.A. Whitmore (Editors), Case Studies in Data Analysis. viii, 262 pages, 1994.

Vol. 95: Shelemyahu Zacks, Stochastic Visibility in Random Fields. x, 175 pages, 1994.

Vol. 96: Ibrahim Rahimov, Random Sums and Branching Stochastic Processes. viii, 195 pages, 1995.

Vol. 97: R. Szekli, Stochastic Ordering and Dependence in Applied Probablility, viii, 194 pages, 1995.

Vol. 98: Philippe Barbe and Patrice Bertail, The Weighted Bootstrap, viii, 230 pages, 1995.

Vol. 99: C.C. Heyde (Editor), Branching Processes: Proceedings of the First World Congress, viii, 185 pages, 1995.

Vol. 100: Wlodzimierz Bryc, The Normal Distribution: Characterizations with Applications, viii, 139 pages, 1995

Made in the USA
Monee, IL
07 July 2026

56544410R00085